BASICS OF FAST ELECTRIC R/C BOATS

by JAY TURNER

RADIO CONTROL BOAT MODELER

DEDICATION

It is customary to dedicate a book to one's parents, spouse, or good friend, and I won't deviate from that tradition. This book is therefore dedicated to my wife, Karen, who has been both a long-suffering victim of the R/C boat hobby and a willing, excited participant. I thank her for her patience, her help and her inspiration, on this project and on many others over the years. We make a good team.

BASICS OF FAST ELECTRIC R/C BOATS

Group Publisher: Louis V. DeFrancesco, Jr.
Group Editor-in-Chief: Tom Atwood
Technical Editor: Gerry Yarrish
Copy Director: Lynne Sewell
Managing Art Director: Alan J. Palermo
Director of Marketing: Gary Dolzall
Production Manager: Mary Reid McElwee

ISBN: 0-911295-35-6
 Published by Air Age Inc., 251 Danbury Rd., Wilton, CT 06897. Dr. L.V. DeFrancesco, Chairman of the Board; M. F. Doyle, Chief Executive Officer; Y. DeFrancesco, Publisher.

PRINTED IN THE USA

ABOUT THE AUTHOR

Author Jay Turner began to run R/C electric boats in the early 1960s. The first R/C model he ever saw was a gas airplane; it fascinated him until it suddenly dove and was reduced to a pile of many small pieces on the ground. He immediately decided that too much work was required to build a model aircraft that could be destroyed so easily. Boats seemed safer, and he loved all things "marine," so starting his R/C career with a model boat seemed a natural path to take.

That first model was a Dumas PT 109 powered by a 6V motor and a wet-cell battery—state of the art at that time. The simple Controlaire single-channel radio with a Boatomatic combination rudder and motor controller worked well, and Jay enjoyed both the look of this "fast" electric model and its scale-like speeds. He was eventually seduced by the greater speed and excitement of glow-powered models, and over the next few decades, a series of noisy deep-vees, tunnels and hydros passed through his hands—interspersed with a few electric specimens.

In the mid-'80s, the fast electric boat phenomenon was growing rapidly, and discouraged by a lack of good running sites for his loud glow models, Jay decided to get back into electric boats in a big way. With several friends, he formed the Anchorage Model Boat Club (one of the most active fast electric clubs in the country), and he has continued his interest in the hobby, despite the rather isolated nature of the club (the nearest electric club is a three-hour plane ride away!).

Though he lives in Alaska, he keeps his finger on the pulse of the hobby by running in national fast electric events, and he is in close contact with the directors of the national model electric boat racing organizations. In 1994, he was appointed assistant director of fast electrics for the International Model Power Boat Association (IMPBA)—an appointment he was very honored to receive.

Jay continues to experiment with all aspects of fast electric boating, including the new drag-racing venue. He writes for *Radio Control Boat Modeler* magazine—articles on boat construction, drive systems and other technical aspects of the hobby. He continues to build on his knowledge of fast electric boats and its ever changing technology, and he can't foresee the day when he will no longer be excited by racing them.

ACKNOWLEDGMENTS

No book of this scope can be written without the help of a number of individuals. Though I can't mention everyone, I give special thanks to the following:

Fast-electric directors Jim Dunham (International Power Boat Association) and Dick Crowe (American Power Boat Association) for their assistance and expertise on a wide range of technical matters.

Fast R/C boat guru Ed Hughey (Hughey Boats) for sharing his time and experience of boat design, drive systems and setup.

Motor makers Bob Boucher (Astro Flight), Neal McCurdy (Trinity) and Mike Walker (Twister Motors) for their patience over the years with my questions about their own products and electric motors in general.

Editor Gerry Yarrish and the crew at Air Age Publishing for their guidance and help in gathering the materials needed for the finished product. The professional look of this book is due to their experience and expertise.

Finally, the members of the Anchorage Model Boat Club for their support and help with questions, their enthusiasm for fast electric race boats and their allowing their models to be photographed. *—Jay Turner*

C O N T

ENTS

INTRODUCTION

TRULY FAST electric model boats are a relatively recent phenomenon. Twenty-five years ago, electric record holders barely topped 20mph when running on over 24 volts, and they often burned up the automotive windshield wiper or starter motors used to achieve those speeds. As well as that, the heavy wet-cell batteries they ran required overnight charging, and that meant that a boater could get only one run per day per battery. No wonder "fast" electric boats were never popular.

What really got fast electrics off the ground was the R/C car hobby. Suddenly, there was a huge market for both high-performance and high-volume production, and the result was a quantum leap in electric car technology, especially in motors, radios and batteries. As more and more car racers got into the hobby and bought more cars and accessories, the price of almost everything associated with R/C cars decreased, so the hobby became even more affordable. A few model boaters recognized the possibilities of the new technology for model boats and began experimenting to increase both speed and enjoyment.

In the late 1980s, some of these boaters published the results of their efforts, and manufacturers started to offer boats and equipment especially designed for marine use. Back then, an electric hydro that reached 20mph on seven cells was considered extremely fast; today, 7-cell records top 40mph. If you know how, it's now fairly easy to assemble the parts to create a fast, dependable model boat. I hope that this book will help to point you in the right direction to do just that.

CHAPTER 1

Hull Types

1 HULL TYPES

This 1/16-scale unlimited hydro looks and performs like the full-size version. Scale models have a lot of appeal, and fast electric scale hydros are almost as fast as outriggers.

MONOS: DESCRIPTION AND THEORY

Monoplanes have a single planing surface. Common examples are the pleasure boats and fishing boats that make up most of the private craft in this country. Though a small rowboat, a large cabin cruiser and a powerful offshore racer may look different and perform different duties, they are all classed as monoplanes because they have just one bottom surface.

Fast electric model monoplanes fit into two groups: deep-vees and shallow-vees.

• **Deep-vees.** Probably the most common, these are usually the best choice of fast electric monoplanes. With a vee angle of more than 24 degrees, the deep-vee handles rough water better than a shallow-vee because its sharper angle slices through the waves more efficiently.

• **Shallow-vees.** These hulls are often faster in the straights than deep-vees because the flatter bottom rides higher in the water and so encounters less friction with it. Because of its flatter bottom, a shallow-vee will pound the waves badly in a chop and will be prone to taking off when run into the wind.

Cornering a shallow-vee can be exciting because it will bounce and hop around.

Because it has a larger wetted surface area, a deep-vee, on the other hand, is usually slightly slower along the straights. It does "pound" less than a shallow-vee, but its chief advantage for racing is in the turns. While a shallow-vee hops and skips around a turn, a deep-vee will lean smoothly into it and, if set up correctly, will make a quick, easy, high-speed exit onto the straight. In a race, a driver doesn't have to worry about his deep-vee flipping over in a turn, so he can concentrate on strategy and victory.

Good examples of the two mono types are the Graupner* Electro Vee

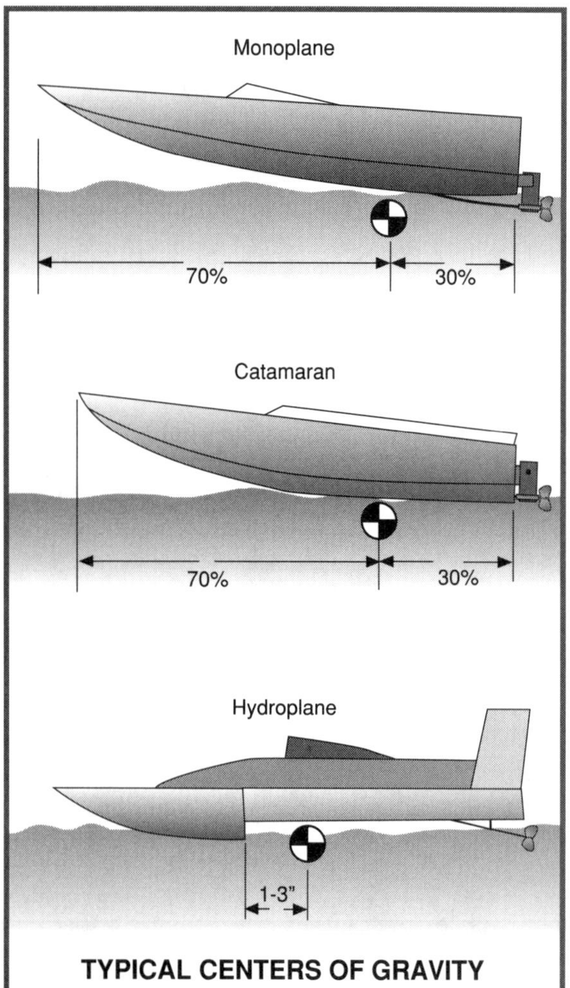

Typical center of gravity (CG) values for various types of hull. These figures are merely guides; your boat may perform best with the CG in front of or behind the positions shown.

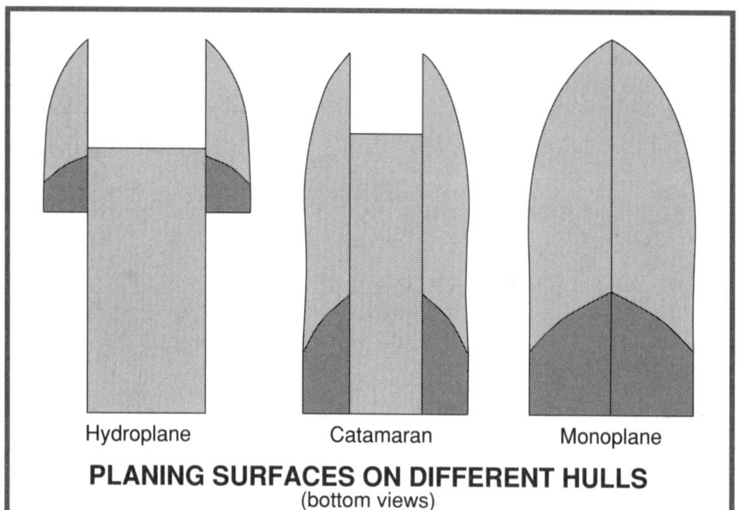

The size of a hull's contact area with the water will affect its speed. The larger the contact area, the greater the drag, and the lower the speed potential. The three-point hydro has the smallest contact area and, thus, the greatest potential for speed. The catamaran is partially supported by air pressure inside the center tunnel, and its contact area is greater than the hydro's. The monoplane has the largest contact area, so its potential speed is less than that of the hydroplane. Note that a very well-set-up mono will run faster than a mediocre hydro.

and the DPI* Crackerbox. The Electro Vee is a classic deep-vee and is great for sport and racing. The Crackerbox has a very shallow vee bottom, so it bounces along the straights and hops through the turns. In anything but calm water, the Crackerbox isn't a good race boat, *except* when racing against other Crackerboxes. When all boats have the same, relatively flat, bottom, the racing is some of the most interesting and exciting, both to watch and compete in.

In turns, a typical deep-vee hull has an advantage over a shallow-vee, but that advantage can disappear if the deep-vee is equipped with a steerable outdrive. In tight turns, the changing thrust angle can make a narrow, deep-vee hull prone to hooking, especially if the boat has been modified to run faster than it was designed to go.

The Kyosho* JetStream and Horizon Hobby* Baltek boats are great entry-level monos, and their steerable outdrives make them very maneuverable. But add a few cells and hot motors so that their speed increases by 50 percent, and they can become flighty and more difficult to drive. This isn't a criticism of these fine models; as designed, they run really well. But be warned: if you modify a boat to increase its performance, you may discover that you have caused problems in other areas.

Examples of the three hull types discussed in the text are the three-point hydro (rear), the monoplane (center) and the catamaran (front). They're about the same size, but the cat and the three-pointer are significantly faster than the shallow-vee monoplane. Still, monos appeal to many modelers, and they can be set up to give a very exciting performance.

HOW TO SET UP YOUR BOAT

It's usually easier to set up a monoplane than to fine-tune a hydroplane, but if maximizing your boat's performance is your goal, there are certain rules to follow.

• **Weight distribution** is very important for high speeds and good handling, and finding the model's balance point is the first step in fine-tuning the hull. Multiply the hull's length by 0.3, measure this distance forward from the transom, and mark that point on the deck. Check the model's actual balance point, and then move the batteries fore or aft to make the boat balance on your mark; then test run it to see how it performs. If the bow rides low and plows water, move the balance point back an inch; if the bow hops up and down, move the batteries forward. You'll probably see the fastest speeds when you've trimmed the model so that its bow is just a little light (showing a very slight oscillation while running).

• **Propellers** can also affect a mono's running attitude. Generally, low-lift props such as those in Octura's* X-Series will perform best, because they let the stern settle into the water a little and improve cornering and stability. High-lift props such as those in Octura's 1600 Series will raise the stern and lower the bow, and they can make the boat hook in the turns.

Monos have either surface or sub-surface propellers. The difference is pretty obvious: sub-surface props run below the water's surface; surface props run partially out of the water. Both types offer advantages, but today, most fast electric monos run surface props.

The sub-surface setup gives better acceleration from low speeds because the prop gets a good "bite" on the water as soon as the throttle is

Two winning hull designs from Graupner—the Monster Vee monoplane and the Tornado hydro. The owner of these boats has hulls set up for several racing classes, and he can run his four boats in at least eight different classes.

Two pure, racing, outrigger hydroplanes: left—a conventional outrigger; right—an experimental design.

advanced. It works well at lower speeds, and for scale-like boats it may be preferable to a surface drive.

A surface drive puts the prop partially out of the water when the boat is on plane, and this increases speed because it reduces drag. Also, because the prop is pumping air for part of each revolution, there is less load on the motor, so prop rpm will increase. This means that a larger propeller may be used, and this causes an additional speed increase. Disadvantages to the surface drive are more sensitivity to trim and fine-tuning, the need to use surface propellers, and poor low-speed performance and acceleration. Still, the surface drive is most popular for its higher speed potential and the impressive, exciting rooster-tail it produces.

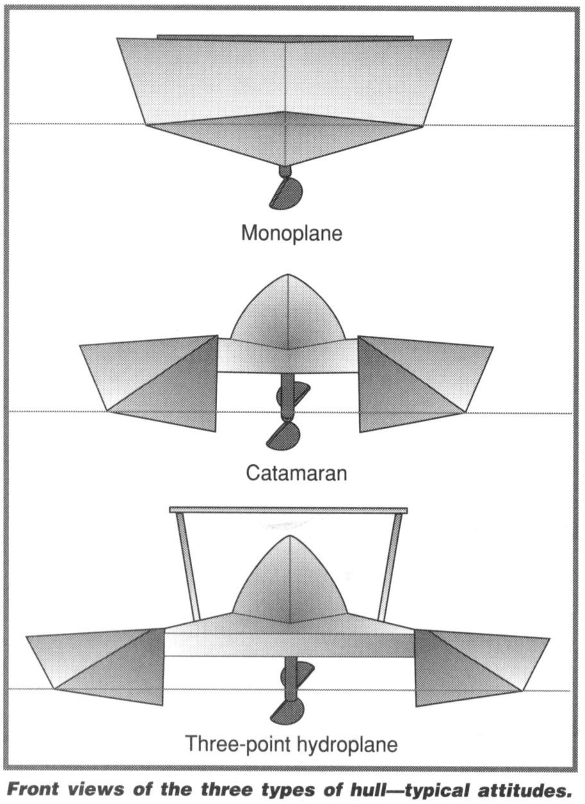

Front views of the three types of hull—typical attitudes. Note that the catamaran and three-point hydro use surface-drive propellers, but the mono's prop is fully submerged. Most cats and hydros use surface drives; monos can use either to good effect, although the most recent trend is toward the surface-drive setup.

The fastest monos run with only the last few inches of the bottom in the water, but running like this reduces stability and makes turning difficult. In the turns, you may have to slow down quite a lot to avoid flipping or spinning out. Depending on the water conditions and the boats you run against, sacrificing some top-end speed may actually lower lap times because you won't have to slow down so much for the turns. Try to trim your mono to achieve a fast but stable speed; no matter how fast your boat is in a race, if it flips, you lose!

What is Hooking? How can it be Avoided?

If a model "hooks" in turns, its bow catches in the water and the boat pivots around the prop and spins out. If the stern lifts in turns, this hooking tendency is magnified, so keeping the transom down becomes important. To eliminate hooking, install a spin fin on the *right* corner of the transom; this will prevent the stern from swinging around in tight right turns. The spin fin should be installed at a right angle to the boat's bottom. Depending on their design and setup, some boats will benefit from having a smaller fin added to the *left* corner of the transom (in addition to the right spin fin). Experiment with fins of several sizes, and use the smallest fin that will eliminate the hooking (large fins cause drag and slow the boat).

HYDROS: DESCRIPTION AND THEORY

Hydroplanes are the fastest, most exciting electric boats, but they are also the most difficult to trim. A hydro is fast because very little of its hull is in the water (low drag), but this means that there is very little contact with the water to keep it stable. It isn't unusual to see fast electric hydroplanes flip in a race, or even when running alone if there's much wind or chop on the water. Trimming a hydro involves a compromise: high speeds (minimum hull/water contact) must be balanced against having enough stability to stay upright under racing conditions.

The two most common hydro hulls

are the catamaran and the three-point. The two types behave differently, so they will be discussed separately.

Two outboard tunnelhulls exit a turn together. A derivative of the catamaran, the tunnnelhull uses an outboard drive rather than an inboard one.

• **Catamarans** (cats) are, in essence, two parallel hulls separated by a tunnel. When the cat is traveling at speed, the tunnel is filled with air and helps to lift the boat out of the water with air pressure. The two sponsons provide good stability, and the long contact area makes turning sharp and well-controlled. As in monos, the center of gravity will be about 30 percent forward of the transom, and also like monos, the cats are pretty stable and easy to drive. They can be very fast, too, and if set up to "fly" down the straights with a bow-high attitude, a cat will run almost as fast as a good three-point hydro.

The advantage of the catamaran is its superior turning ability; unlike a three-pointer, it won't slide in the turns, and by driving a tighter line around the buoys, a cat can keep ahead of a faster hydro that drifts wide.

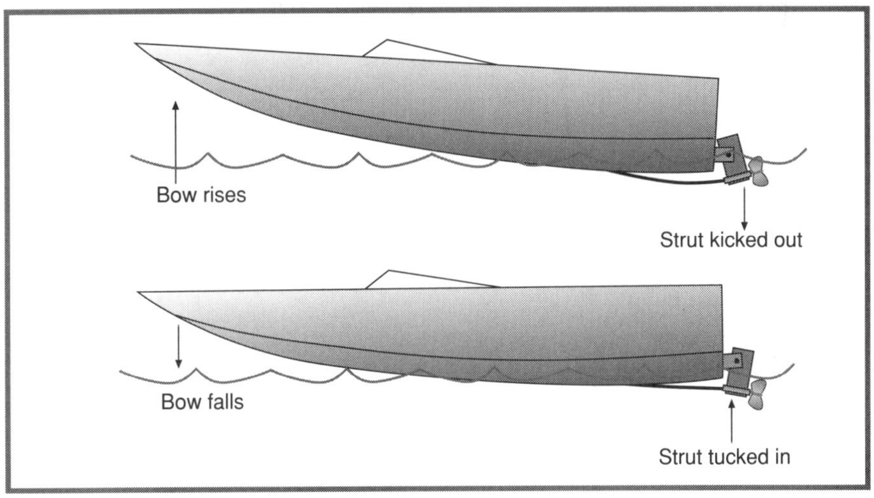

Adjusting the strut can affect a model's ride attitude and, thus, its speed and handling. Kicking the strut out pulls the transom down and raises the bow; this sometimes increases speed. Tucking the strut in will lower the bow and might improve turning. Too much tuck-in can cause the hull to hook and spin out.

• **Three-points** are the fastest type of model boat. A typical three-point hydro has two sponsons in the forward half of the hull to support the boat's weight, and in the stern, the propeller acts as a lifting surface to hold up the transom. These three small points of contact give these boats their name, and the small contact area is the reason for their high speeds.

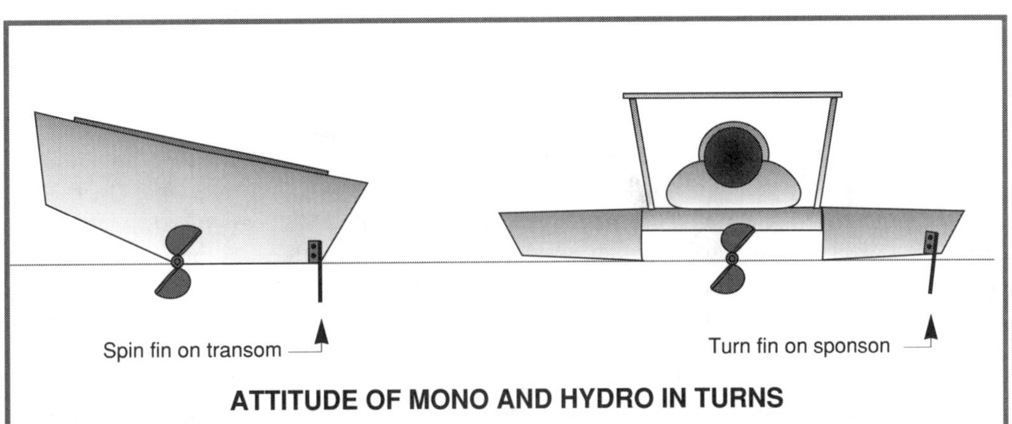

A well-set-up mono will bank into the turns, and any tendency to spin out will be checked by a spin fin. The hydro maintains a flat attitude, and it desperately needs the turn fin to hold it in the turn.

Common three-point hydroplane types:
— scale three-pointer;
—outrigger.

Although they're both hydros, they have different racing potential owing to their design. The scale boat looks like a full-size hydro and appeals to many modelers. Their performances can be very exciting, especially when 1/16-scale

This single stepped hydro differs slightly from the traditional three-pointer, but like that hull, it can reach very high speeds owing to its having only minimal contact with the water. As this photo shows, only the rudder and prop are in the water at all times.

hydros on seven cells and in oval racing trim hit well over 30mph. Because their design traps air underneath them, they are more prone to "taking off" than other hull types.

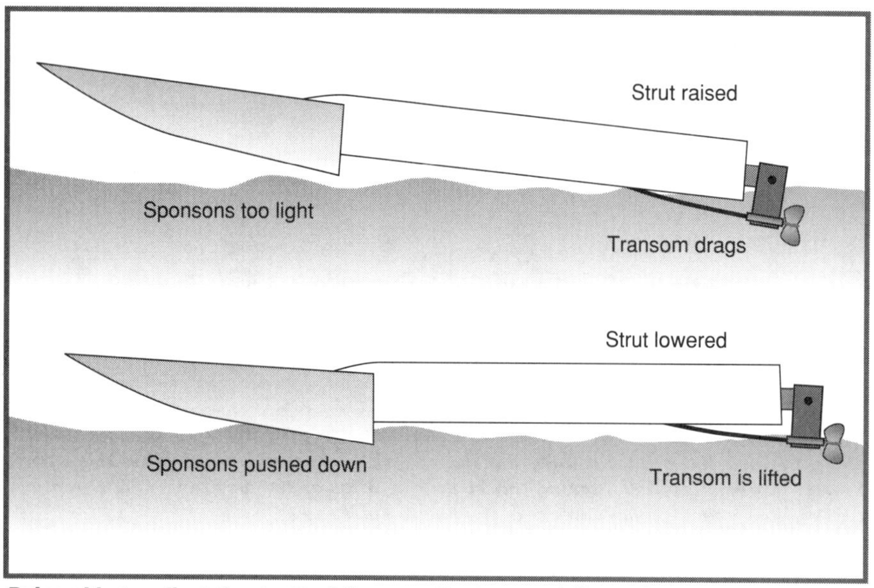

Being able to adjust the depth of the propeller is especially useful on hydros. On catamarans, raising the strut may help top speed by allowing the prop to unload, but on three-point hydros, this could cause the transom to drag in the water and significantly reduce speed. Lowering the strut will help to raise the transom out of the water, thereby reducing drag and increasing speed.

Because it has less air trapped under its hull, the outrigger hydro design has none of the "flighty" nature of the scale hydro. These "stick" boats don't look like "real" boats, but they perform better than any other model electric boats. They weigh less than scale boats and seem to run faster, too. Even with these supposed advantages, many oval racing records are held by scale hydros rather than by outriggers.

With hydros running at speeds of more than 25 to 30mph, instability and difficulties when turning can be problems, but to the avid hydro driver, the boat's exciting speed potential makes these problems worthwhile. Unlike a cat, which will "hook up" well in a turn, a typical three-pointer will slide and hop through turns because its small turn fin prevents it from plowing straight ahead. With a center of gravity close to the midpoint, if a hydro gets its nose up too high, it will "kite" off the water and fly for a short distance. The landing isn't usually pretty!

CATAMARAN SETUP

Cats are not overly difficult to set up, especially when compared with the typical three-pointer. Start with the center of gravity placed 30 percent of the boat's length ahead of the transom, and use a low-lift prop, such as those in the Octura X Series. Position the prop shaft so that it's in line with the bottom of the sponsons to start, and move it upward or downward until the speed tops out.

On occasion, a cat will not "air out," and the tunnel will stay filled with water, which produces enormous drag. This is the result of having too little power, too much weight, a poor thrust angle, or the wrong type of prop. Another reason is lousy hull design, but most cats can be made to run well with some tuning. First, try moving the center of gravity back to lift the bow; next, try adjusting the prop angle by kicking back the prop. Make sure the motor is in good shape; you may have to upgrade a kit's motor just to get the model to run properly. You may also have to experiment with props of various sizes and types.

Assuming the boat airs out and runs reasonably free, what else can you do to increase speed? Again, motor, battery and prop upgrades will usually lead to a substantial increase in speed, but a few other tricks may help, too. Raise the prop 1⁄16 inch at a time until it just begins to cavitate too much, then replace it with a propeller that's the next size up. This may be enough to "hook up" the

An 8-cell mono exits a turn. This gear-drive model rides well out of the water, yet still has enough contact with it to stay under good control.

prop, and by running the larger prop higher out of the water, you'll achieve higher top speeds.

Another modification worth trying is fitting small steps to the bottom of the sponsons about 50 percent ahead of the transom. Try temporarily gluing ⅛-inch-high steps to the sponsons with a drop of silicone glue. The steps may help break up the water under the sponsons to reduce friction and drag; they may also help to lift the bow and trap more air under the tunnel.

This three-point hydro is just leaving the shore. Note that there's no rooster-tail, and the transom is still dragging slightly.

THREE-POINT HYDROPLANE SETUP

Even if poorly set up, cats and monos will usually at least get up on plane, but a three-point hydro may not. A hydro's limited planing area may not be enough to get the boat up on top of the water without some tuning. Before you even run your hydro, make certain that the motor and drive train are free and functioning perfectly. Set the center of gravity within 2 inches of the rear edge of the turn fin, install a proven prop, and set the model in the water. Give the boat half throttle and see how it behaves. If it doesn't look as if it will get up on plane, get off the throttle, then suddenly give it full throttle. At this point, failure to plane may mean that other work is required (see Chapter 11 on "Troubleshooting").

If the hydro gets up on plane, run it around for a while and carefully study how it rides. The transom should run at least slightly out of the water; if it drags on the water in the straights, the boat will be too slow. If the model has a flex-cable drive, lower the prop ⅛ inch into the water and run the boat again. Increasing the prop angle (pointing down at the stern) will also help to lift the transom. If your boat has a solid drive shaft, lift the stern by moving the prop backward (use a longer shaft if you need to). Octura's X Series props work well on most hydros, but you should also try a high-lift prop such as those in the 1600 and 1700 Series. Another trick to try is moving the center of gravity forward to take weight off the prop. The goal is to get the transom clear of the water as soon as possible after the boat exits a turn to minimize drag and maximize speed.

Running with the transom too high can slow the boat by making the front sponsons dig in too much. Lowering the prop or running it at a lower angle will take some weight off the front sponsons and make the hull tiptoe across the pond.

All hydros slide somewhat through the turns, and this sliding is limited by the turn fin. Because the turn fin induces drag that can slow the boat, use the smallest one that will allow the model to turn properly. Turning performance can often be improved by moving the center of gravity backward, but don't move it farther than 2 inches aft of the turn fin's rear edge.

If a hydro with a proven design just won't run properly, there's probably a problem with your motor, batteries, or the way they have been installed. Hydros designed for certain drive systems may not perform well with others; for example, a hull that was designed to use a flex drive may not run with a solid shaft until you've done lots of work, including re-mounting the prop shaft at a different angle.

The top surface of a hydro can affect its performance, particularly in poor weather. Many scale models have rear wings like those on their full-scale counterparts, but the wings increase aerodynamic drag and, under windy conditions, can cause handling problems. On slower hydros, the speed penalty is usually small, but as speed increases, the drag caused by a wing does, too. If you don't need a wing for scale effect, leave it off. In a few cases, the wing will help to lift

the transom and will increase speed. I designed one hydro that had a large rear wing, and when the boat's speed went above about 25mph, it did a good job of lifting the transom. Because the model was designed around the wing, it simply did not work without it.

Some boats benefit from front spoilers, which keep the bow on the water at the end of a straight. I've seen more than one super-fast 7-cell scale hydro take off after catching a slight breeze at the end of a straight-away when its speed was highest. Fitting small spoilers or wings near the bow can help to keep the boat on the water.

You'll find additional information and tips on the many available hull types in John Finch's book, "Advanced R/C Boat Modeling," (published by Air Age Publishing*). Though John deals exclusively with glow-powered boats, these and fast electric models have much in common. I encourage you to buy a copy of John's book and expand your knowledge of R/C boating.

CHAPTER 2

Drive Systems

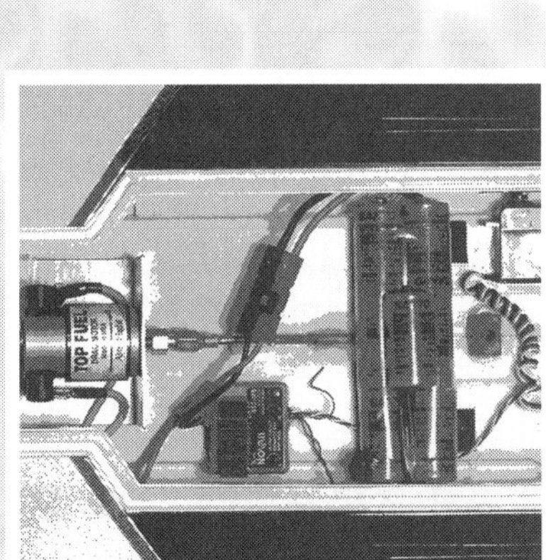

GEAR DRIVES

Glow-powered boats are almost exclusively direct-drive models; a shaft connects the engine with the propeller. This is simple and efficient and, in most cases, it's the best way to transfer the engine's power to the prop. If more thrust is required, a larger engine/propeller combination is chosen.

With fast electric models, the situation is different. Here, many boats use 05-size motors, and if additional power is needed, either a more powerful motor or more batteries will be installed. In most cases, though, these changes will require that a prop with a smaller diameter be used to avoid overloading the motor. The hope is that the smaller prop will be spun faster and, ultimately, will provide more speed.

A problem develops when a larger hull is used. Assume for a moment that a 36-inch monohull is to be powered with a single stock 05 motor on seven cells. A 35mm-diameter prop is used, but the large hull is too heavy and has too much resistance to get up on plane. A more powerful modified motor is installed along with an 8-cell battery pack, but the 35mm prop is too large for this combination and overloads the motor. Changing to a 30mm prop relieves the motor, but the new thrust cone is too small to push the boat, and it goes more slowly than before.

The solution is to use a larger propeller with its larger thrust cone. But because a typical, single, direct-drive 05 motor will be overloaded with a larger prop (bigger than 30mm diameter), you must use either a larger motor (Astro Flight* and Graupner are two suppliers of larger motors) or a gear drive. Most current model manufacturers know this, and large monohulls such as the Hobby Dynamics CESA and the Kyosho JetStream come with gear drives and larger props. These models provide good entry-level performance and, with a little hopping up, can be made more competitive for fast electric racing.

Gear drives installed in most factory boats lack durability when pushed outside their design limits. Though they work well with the mild stock motors supplied, they sometimes fail when more potent motors and batteries are teamed with larger propellers. The increased stress can be too much for the stock gears, and they can break when the boat is raced or driven hard. The solution is to use a high-quality aftermarket gear drive by Hughey Boats*. These products are designed for stressful, high-performance use, and if well maintained, they'll provide years of trouble-free running. Available as both single- and dual-motor units, the Hughey products are the "gold standard" for aftermarket gear drives.

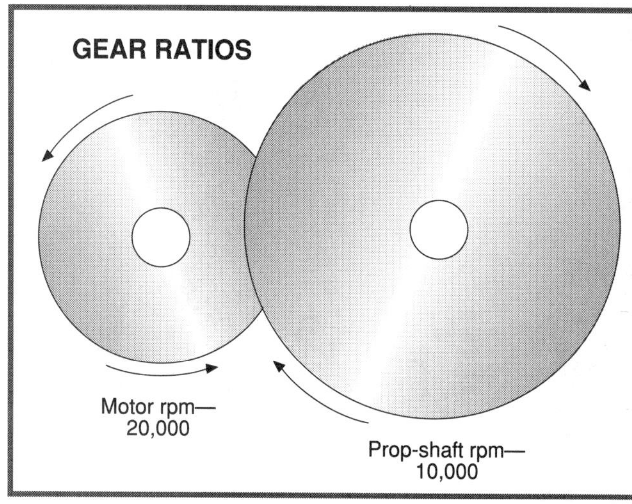

Gear drives reduce the strain on the motor when large props are used. In this illustration, the motor is allowed to turn at twice the speed of a large prop. This reduces amp draw at the motor yet supplies enough power to spin the big prop.

WHY RUN A GEAR DRIVE?

Simply put, a gear drive's advantage lies in allowing a motor or motors to turn faster than the propeller, drawing fewer amps than a direct drive and reducing heat, wear and tear. For fast boats, typical gear ratios are 1.5 to 2.5:1. A general rule is that the larger the propeller or the fewer the turns of wire on the motor, the higher the reduction must be. The 36-inch mono in the above example might run a 42mm propeller with a stock motor and a 1.8:1 gear ratio. For higher performance, the same prop with a 15-turn motor and 2.25:1 ratio might be used.

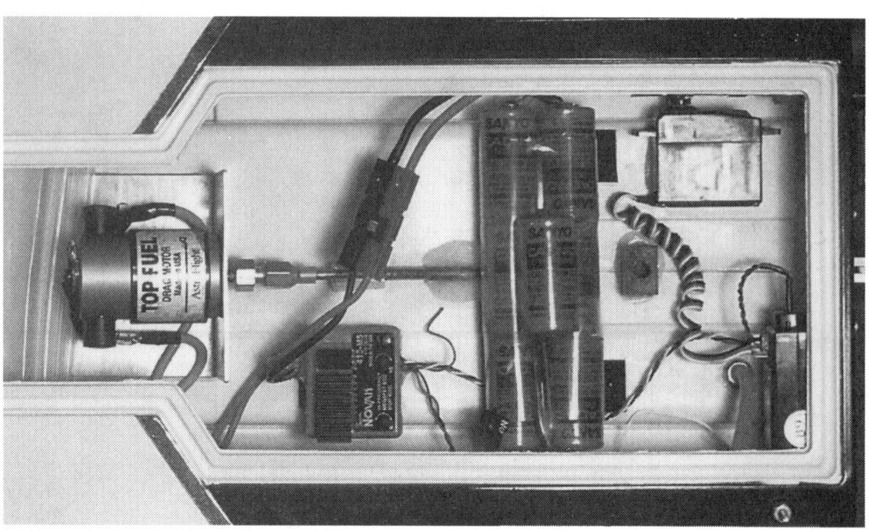

Here's a typical direct-drive setup. The motor is mounted on an aluminum plate and fitted with an Octura flex-hex coupler. The drive cable is clamped in the coupler and runs through a stuffing box to the strut.

An additional but related use for gear drives is with high cell counts. If correctly propped, most 05 motors can be used with up to eight cells without self-destructing. If you want to run in 12-cell events , you should use either a special 12-cell motor or a dual-motor gear drive. The dual drive uses two 05 motors, usually wired in parallel. Set up this way, each motor is supplied with maximum voltage, but with only half the available amperes (amps). This means that the same amount of power (watts) comes out of the drive unit, but since the motors each use only half the available amps, they don't overheat, and their run times are extended.

Just installing a gear-drive system in a direct-drive boat without making any other changes will result in longer run times because the motor will be turning faster and using fewer amps. This is all well and good for scale boaters who want to cruise sedately around the pond, but fast electric boaters will sacrifice run time for speed. To effectively use the available battery power, the motor must be more heavily loaded and must draw more amps, and this is done by running a prop with a larger diameter and/or greater pitch, and by using a motor with fewer turns. The desired result is an increase in the size of the propeller thrust cone, a reduction in motor rpm and an increase in boat speed. The limit is reached when run time is insufficient to complete a race, or the motor overheats.

In organized racing, an additional advantage of having a gear drive becomes apparent. Many two-day events allow boaters to run in both stock motor and open (modified) motor races on alternate days, and being able to run one hull in both events gives you an obvious cost advantage. If a stock 05 motor must be used, it can't be run direct drive in 12-cell classes without the risk of damage caused by overheating. By bolting in a dual Hughey drive unit, a racer can run two stock motors in the stock class and then switch to two modified motors for the open classes. Many racers do this at national events, and the twin-motor setup has won plenty of stock-class trophies.

When a specific application requires more than a single direct-drive 05 motor, a special, larger motor can be used in place of a gear drive (described in the next section). But a modeler may be more comfortable with the familiar R/C car motors, and a gear drive allows him to use them. Many racers enjoy experimenting and tinkering with R/C car motors, and with gears, they don't give up any of the advantages of direct drives.

Experimentation is the key to obtaining the best performance with gears. If your boat originally came with a gearbox, vary the gear reduction, and try running with higher or lower ratios. Install a Hughey drive in your next fast boat, and see how it performs. As you test different combinations, you'll find that substituting a more powerful motor or more cells will probably require a slightly smaller prop and/or a higher ratio, but only by testing will you be able to find the best combination.

The Hughey twin-motor drive system is for use in models that use more than eight cells. The gear sizes can be altered to modify the gear-reduction values. Hughey makes a single-motor gear drive, and it, too, is a popular accessory for boats that run with six to eight cells.

DIRECT DRIVES

Gear systems are popular and can be a good solution to some boating problems, but they aren't always needed. Direct drive will often produce the fastest boat, particularly with smaller models running six or seven cells. Direct drives are cheaper, simpler, lighter and more mechanically efficient than gear drives. For many fast electric applications, they make more sense.

A 7-cell hydro or monoplane that's less than 30 inches long and weighs less than 3 pounds is a good candidate for direct drive. Motor selection is fully discussed in its own chapter, but a good place to start is with a 16- to 20-turn modified motor and a 27mm to 32mm propeller.

The smaller and lighter your boat, the faster it will run, so if you want the fastest boat, choose one that's between 22 and 28 inches long. If it weighs less than 3 pounds ready to run, so much the better. Smaller, lighter boats can use more powerful motors and larger props, and this increases speed even further. As with gear drives, experimentation is the only way to find the best combination.

If special motors are installed, direct drive can be used in larger, heavier boats running 12 cells and more. Astro Flight's powerful 326

The Hughey drive is mounted in the hull shown on page 16. The original 8-cell Astro Flight-powered model was fun and reasonably fast, but with the twin drive and 12 cells, its performance is ballistic!

motor runs very fast on 12 cells, and Graupner's Turbo 700 Series motors will work well on 12 cells at a lower unit cost. The same truth applies to these larger boats as to the smaller 7-cell models: lighter is faster.

INBOARD SYSTEMS

This is by far the most popular installation, and it's the one that provides the highest speed potential. The motor is usually mounted about 40 to 50 percent of the hull's length forward of the transom and is connected to the prop by either a solid shaft or a flexible one. Because their output shaft is usually much lower in the hull, gear drives allow the motor to be placed closer to the transom without using an excessive prop-shaft angle.

It's generally much easier to balance the hull with an inboard installation, and if the hull is adequately sealed, the motor and ancillary equipment are kept dry, even if the boat flips over. With the motor sealed in the hull, the heat generated by the motor, speed control and batteries can be a problem, but water cooling the motor (and the speed control, if possible) will limit the heat or prevent it from doing any damage.

Hydroplanes and shallow-vee monos will benefit from having a low center of gravity (CG), so if you have the option, mount the motor and batteries (the heaviest components) as low in the hull as possible. In many hydros, the motor is actually in a motor well below the hull bottom to keep the CG low. When the boat is up on plane, the motor well is above the surface of the water, so it doesn't contribute to drag.

Deep-vees can benefit from having a higher CG that helps them to lean into turns. When a boat holds a steady, banked attitude, its cornering is faster and more stable, and that, after all, is what we're interested in. Don't go overboard here, but mounting the batteries several inches above the keel can help a deep-vee to turn without losing much stability on the straights.

OUTBOARDS

Fast electric outboards have an enthusiastic following, and though they seldom exhibit the raw speed of the inboards, their scale-like appearance and ease of cooling make them attractive. The most common outboard hull is the competition tunnel; the MRP* Bud Light racer is typical of the breed. When trimmed correctly, the tunnels are quite fast down the

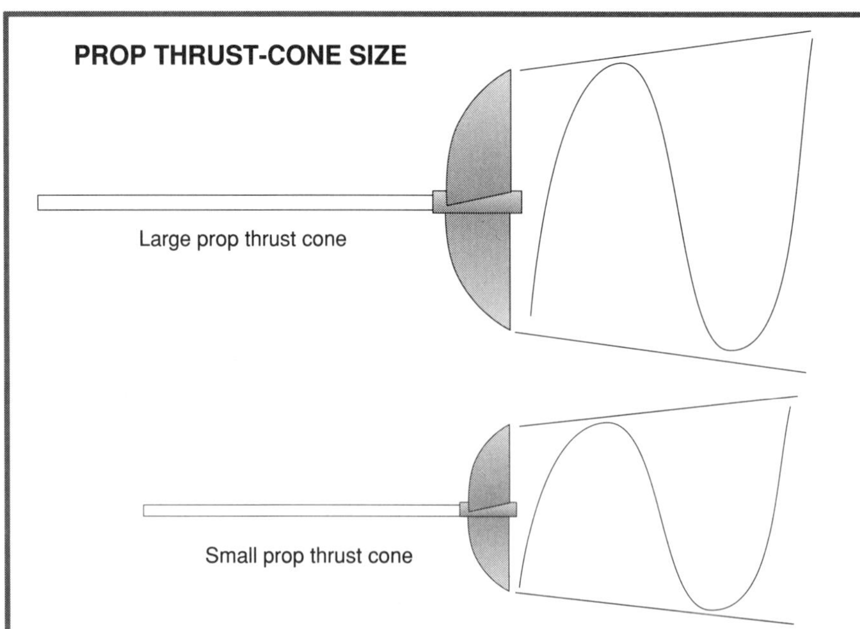

A larger prop produces a larger thrust cone than a small prop, and big, heavy models are pushed more efficiently by a large thrust cone. To efficiently increase a model's thrust cone, use a larger prop turned by a gear drive, or use a larger motor of a different design.

straights, and their turning ability is awesome. They share one undesirable trait with their full-size counterparts, though: if run too fast into a wind, they'll flip over backwards.

The only other widely available type of outboard hull is the catamaran (for example, MRP's Bud Dry). I've seen some very interesting, wooden, outboard hydros by Blackwell Boats*. They look very much like the full-scale APBA B- and C-class hydros, which you'll see running in events all over the country. Outboard monos are certainly possible, and a great many existing inboard hulls could be outboard powered.

Though inboards are by far the most popular type of drive system, outboards are making inroads, chiefly because of the availability of the MRP outboard. This scale racing mono has an MRP outboard mounted on the transom, and it looks very much like the full-size racing utility boat it represents.

There are two major trends in fast electric outboards:

• **The direct-drive approach**, e.g., the MRP outboard. Here, an injection-molded lower unit houses a flex shaft and lower bearing; the motor is attached at the top, and the prop is screwed on the bottom. With proper maintenance, this unit works very well and can provide impressive speed.

Several years ago, my wife dominated the 8-cell hydro class with her pink Bud Light and a 15-turn triple hand-wound motor driving an X430 prop. We used a standing start, and the holeshot the deep prop setting gave her put her in the lead by the first buoy. In the straights, her boat gave up a little speed to the inboard hydros, but on the short, tight course we used that year, her hull's turning ability gave her the advantage.

The disadvantages of an outboard are real, and the modeler who wants to try one should understand the problems. With the motor mounted behind the transom, it's more difficult to balance some hulls properly, and in a few cases, it may actually be necessary to add weight at the bow to obtain the correct balance point. Adding weight is *not* a good way to increase speed.

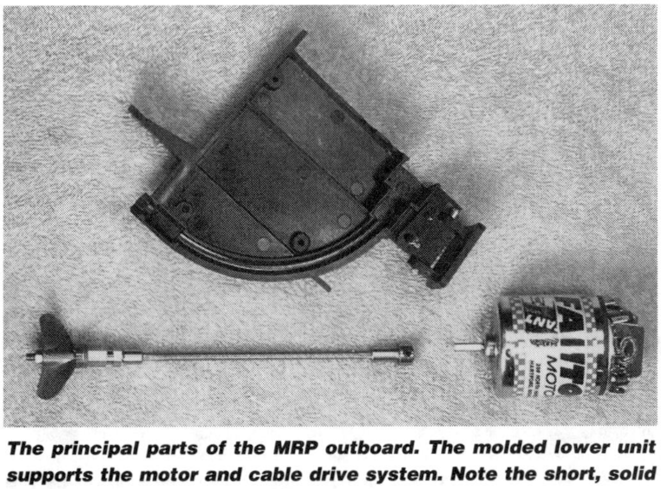

The principal parts of the MRP outboard. The molded lower unit supports the motor and cable drive system. Note the short, solid shaft and bearing block just in front of the propeller; the flex cable transfers power from the motor. An outboard is easier to install than a complete inboard drive.

Some hulls don't respond well to the change in thrust angle when an outboard turns. This isn't a problem on models designed for outboards, of course, but on hulls that were designed for inboard drives, it could cause difficulties—squirrelly handling and hooking in turns. These bad handling characteristics may not become apparent unless the model is traveling very fast, perhaps well over 20mph.

When the motor hangs off the transom, it's exposed to both spray coming off the model and to other boats' rooster-tails—and the occasional flip. I learned the hard way that, for top performance and long life, you must completely disassemble the outboard after every outing. Of course, this is necessary with inboards, too, but with an outboard, the procedure is a little more involved.

The lower unit's halves must be separated and the flex cable removed and dried off. Dry the small drive shaft and check the prop for damage. The motor must be taken off the lower unit and carefully dried; spray it with motor cleaner, check the brushes for wear, and be sure to oil the bearings. I usually keep the outboard parts together, but unassembled, in a zip-lock bag until I want to run the model again.

With practice, it takes just a few minutes to reassemble the unit and mount it on the model's transom. Lube the flex cable with waterproof cable grease, and when assembling the lower unit, be very careful to ensure that the cable ends are properly mated with the drive shaft and the motor connector. The prop nut must be tight, or the drive shaft may become loose when the model is out on the water. A drop of oil on each bearing, and the motor is ready to be mounted back on the transom.

If it sounds as though I'm trying to discourage you from using an outboard, I'm not; nothing could be further from the truth. I enjoy running and watching my wife's tunnel, and the little extra work required isn't too high a price to pay for the enjoyment of running a fast, yet scale-like model. I also like the ease of the initial installation compared to that of the typical inboard setup.

• **The geared approach.** I have no direct experience with the gear-drive outboards, but their advantages are the same as those of gear-driven inboards: larger props and higher motor rpm. Many units use the lower end of the K&B* 3.5cc glow outboard with the motor and gears mounted on top. The K&B unit uses a flex cable much like the MRP unit's, and setup and maintenance are similar for both types. The plastic MRP lower unit is also used as a base for gear installations.

Trimming an outboard can be more involved than trimming the typical inboard, but that's because the outboard has more trimming possibilities:

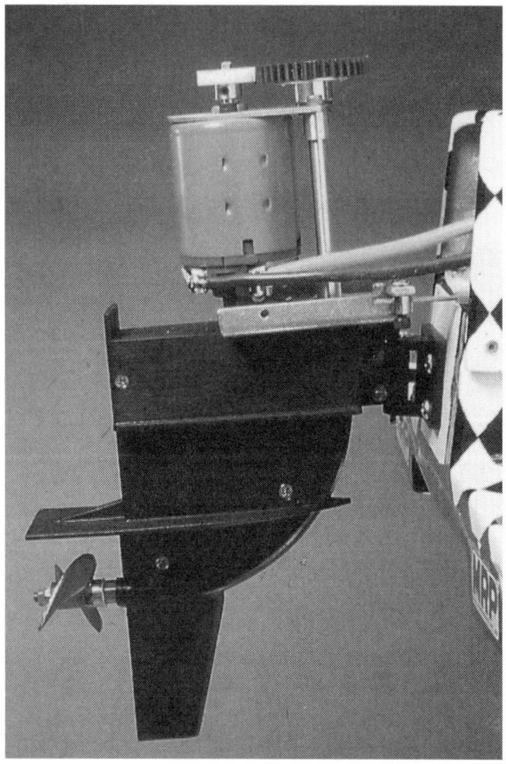

Here, an MRP unit has been modified with a gear reduction to allow larger propellers to be run without overloading the motor. Note the adjustable motor mount and the simple control attachments.

• **Prop depth** is adjusted to give the best speed without compromising the model's turning performance.

• **Drive angle** is varied to tune both the ride along the straights and the turning characteristics. You may kick the prop out to raise the bow and make the model "fly" more effectively, but don't kick it out too much or you'll experience prop blow-out (cavitation), or make the model flip over onto its back when running into the wind. A bow-high attitude is fastest, but turning ability will suffer.

On some hulls, kicking the prop out will also pull the transom down, and that can hurt or help turning (depending on the hull and its speed). Some hulls may need trim tabs to keep the bow down in the straights while they run with the prop kicked out to keep the transom planted in the turns.

The CG should be almost 30 percent of the hull ahead of the transom. This is most easily accomplished by moving the batteries, but a hull's design may limit how far forward they can be moved. Trim tabs or prop downthrust may be needed to keep the bow down and prevent a fly-off. As a last resort, you can add lead weights, but keep them to a minimum, and put them as far forward as possible to maximize their effect.

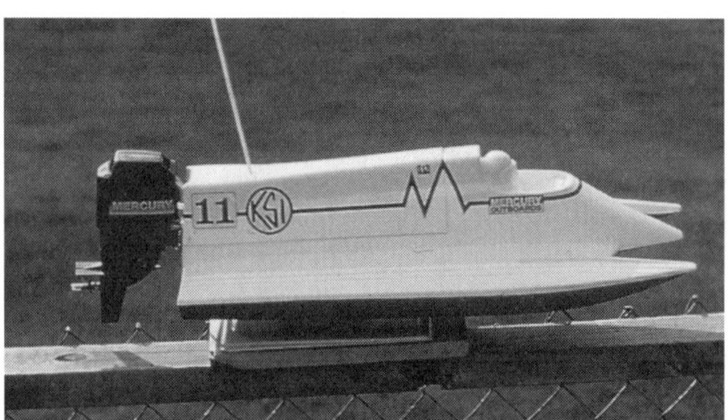

The MRP outboard mounted on the MRP tunnelhull. Originally designed to use 12 cells and an inexpensive motor, this version runs with eight cells and a hand-wound modified motor. Its performance is considerably better than that of the stock version.

This beautiful scale hydro runs with the same drive system as the original—solid shaft, direct drive.

CHAPTER 3

Motors for Fast Electrics

MOTORS FOR FAST ELECTRICS

THEORY AND OPERATION

The electric motors typically used in fast electric R/C boats are virtually identical to those used in electric R/C cars. Most are less than 1.5 inches in diameter and about 2 inches long, excluding the shaft. The diagram on page 24 illustrates the basic parts of a typical motor with brushes; we won't get into a description of the new brushless electric motor that's now beginning to show up in the marketplace.

In operation, an electric motor completes the conversion of the chemical energy stored in the battery into the mechanical energy that turns a propeller. The differences between motors (power and efficiency) are directly related to how a motor was designed and constructed; but battery energy is what ultimately limits the horsepower a motor can produce.

Electric motors produce their power in a different way than gas- or glow-powered engines. The gas engine produces very little torque at low rpm, but as an engine's speed climbs, so does its torque output. At some specific rpm—depending on an engine's design—torque peaks and begins to drop off. Horsepower is developed similarly, and it reaches a peak at higher rpm than the maximum torque. (Torque is the twisting force applied to the prop; horsepower is the amount of work done. Simply put, torque provides acceleration and horsepower provides top speed.)

An electric motor produces its maximum torque at very low rpm, then, as motor speed rises, torque begins to drop off, and this continues until there isn't enough left to turn the motor any faster. For a motor running without load, this speed is the maximum rpm listed by its manufacturer. The motor's mechanical horsepower rises to a peak and then falls off as rpm increase further. Typically, a motor's maximum power occurs at half its no-load rpm, but running a powerful motor this slowly may overheat and destroy it. A better motor speed is 65 to 80 percent of maximum no-load rpm.

The electrical energy used by a motor increases as the motor's speed decreases. The number of amps drawn by a motor closely parallels the torque that it produces and is highest at very low rpm. As shown in the graph on page 27, a powerful modified motor draws more amps at a higher rpm than a less powerful stock motor does. Though slowing a motor down so that it produces more torque and swings a larger propeller may seem like a good idea, that increased torque means more amps used and more heat generated. Running a motor too slowly may result in better performance for a little while, but that motor will overheat and be damaged if it must draw too many amps for too long. Keep motor rpm high, and if you want to go faster, use a more powerful motor and a smaller propeller.

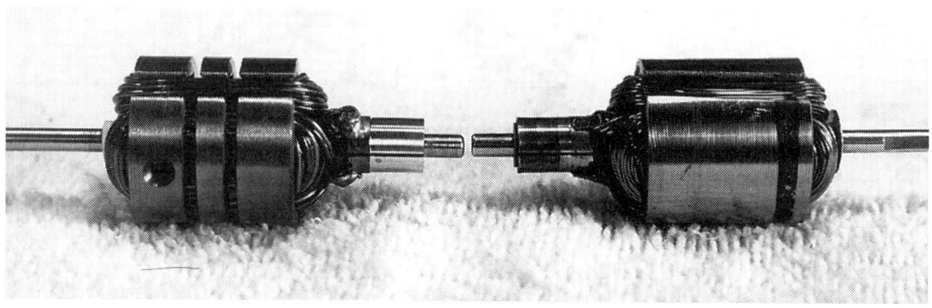

The motor's armature (arm) is the part that spins the drive shaft to turn the propeller. These arms are of a different design, but they do the same thing. The arm on the left is new and has a perfect commutator (on the right end). The comm on the right has darkened because the motor has been run at high amp draws. It can be restored to a like-new appearance by being turned on a comm lathe.

This 05 car motor is attached to an aluminum motor mount with two metric screws. It is shown with a cooling heat sink and a flex-hex coupler attached.

MAGNETS

Today, most electric boat motors use one of three types of magnet:

- **"Wet" magnets** are the most common. They're made of a ferrous (iron) material that's initially shaped as a paste. This is why they're called "wet" magnets; they aren't damp when you buy the motor. Wet magnets produce a strong magnetic field and are resistant to high-temperature damage.

- **"Dry" magnets** have been around longer. They lose their strength when overheated, but the newer, wet magnets hold up better.

Many think that high temperatures reduce the power of a motor's magnets; in truth, it's high current loads that do the *permanent* damage. A magnet will lose strength when it's heated, but its magnetic field usually returns when it has cooled down. High current flows can actually damage the magnets permanently, and this damage is more pronounced at lower magnet temperatures. This causes a dilemma for fast electric boaters; if the motor is water-cooled during a run, the magnets will retain more of their strength, but cooler magnets are more susceptible to being permanently damaged when they're subjected to high amp loads.

- **Cobalt magnets** ("rare-earth" magnets) are most commonly used in the cobalt motors produced by Astro Flight, Trinity* and Graupner. Their magnets have much stronger magnetic fields than the ferrous types and are even more resistant to heat damage. Their disadvantage is much higher cost, but, as we shall see later, that cost may not actually be as high as it at first seems.

All motors should be lubricated regularly, but the oil must be correct for the type of bearings used. Modified motors using ball bearings need a thinner, high-quality oil such as that shown on the left; bushings require a slightly thicker oil such as that shown on the right.

Motor manufacturers are constantly trying to improve their products, often with minor changes like lay-down brushes or fancy endbells, but the major reason for the increase in electric motor power over the past few years has been the constant upgrading of the motor's magnets. Some manufacturers use a numerical version system to identify the newest magnets, similar to that used for computer software. Magnets introduced a few years ago in top-of-the-line motors were called 4.95 version magnets, followed by 5.0 and 5.1 versions. Generally speaking, the newer the magnet, the higher its version number. Frankly, I don't know how much better a motor with 5.1 magnets will perform compared to a 5.0 version, but a 5.2 motor is noticeably stronger than the older 4.95 motors. Expect to pay more for the motor with the most recent magnets, but in a few years, those same magnets will be available in lower-cost motors as the top motors continue to be upgraded.

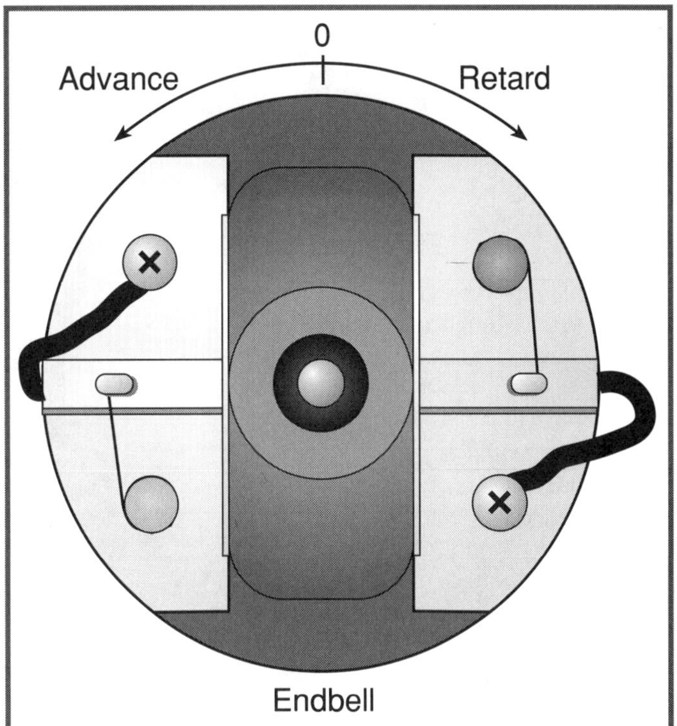

MOTOR TIMING ADJUSTMENT

You can adjust the timing of modified motors by turning the endbell. On an 05 boat motor, neutral or zero timing is a good place to start. Advancing the timing can increase power, but it isn't a good idea to advance R/C car motors by more than 10 degrees (1/8 inch). Under no circumstances should you run a motor that has negative timing—unless that motor is being run in reverse.

ARMATURES AND WINDS

Motors are usually rated according to the number of "turns" on the armature. The armature contains the motor's drive shaft, stacks (or poles) and commutator. Most motors have three stacks on their armatures around which the wire is wrapped. If an armature has a wire wrapped 16 times around each pole, the motor is a 16-turn motor; a motor with 27 wraps is a 27-turn motor, etc.

You might hear a motor called a "16-turn triple." This means that each wrap of wire is made of three thin strands; in a 16-turn single, each wrap is a single, thicker wire. Generally, a motor with a single wind will have more low-speed torque and will draw more current, while a triple or quad wind will be somewhat milder, yet more efficient.

When we compare motors of different types and sizes, all this talk of turns becomes confusing. What really matters is the motor's total resistance, and this depends on the length of the wire that's wrapped around the armature poles. In most cases, the less wire there is on a motor, the lower its internal resistance will be and the greater its power and rpm will be. This means that a 16-turn motor will produce more horsepower at higher rpm than a 27-turn motor does because the 16-turn motor has less wire and, thus, lower resistance.

Some manufacturers have confused things by publishing the "maximum rpm" figures for some of their motors.

There are three problems with "maximum rpm" information:

- The voltage used is often not listed (did they use 5 or 6 volts?).
- The motors produce no useful power at the elevated rpm figure given and usually draw less than 5 amps.
- These figures invite the beginning boater to believe that a motor that produces the highest rpm will be the fastest in his boat—thinking that can lead to nothing but trouble.

Example: a 16-turn motor may be listed with a maximum speed of 26,000rpm. At 5 volts, this motor probably draws about 4 amps at this speed. Four amps is just 20 watts or 0.03hp—not even enough power to push a model boat up onto plane. As a rule of thumb, a typical motor will produce its maximum horsepower at 50 percent of the maximum rpm listed by its manufacturer at the test voltage. This means that as a motor's rpm drop, its amp draw increases and so does its torque. The 16-turn motor in our example should probably be propped to turn around 18,000rpm for good power and reasonable life.

When it's worn, the motor commutator should be re-cut on a comm lathe. This will restore its surface, and the motor will perform like new.

BRUSHES AND SPRINGS

The electricity from the batteries is transferred to and from the armature wire through the motor brushes. Brushes are usually made out of a self-lubricating copper/graphite compound that may be shaped to conform closely to the curve of the commutator's surface. Boats have different power requirements from cars, and unlike cars, the typical fast electric boat pulls maximum power for the entire time it's running. For this use, a "hard" brush is needed to ensure long motor life. Before you break-in your motor, it's best to replace its stock brushes with a hard type.

Soft brushes may produce slightly more power at lower rpm, but the gain is small, and the disadvantages are of more consequence. Soft brushes will wear out quickly and have to be replaced often; brush dust will also build up inside the motor, which will require frequent cleaning. The only racing that might benefit from a soft brush is R/C boat drag racing, in which low-speed torque and acceleration are more important than they are in most oval racing.

The brushes are held in against the commutator by the brush springs,

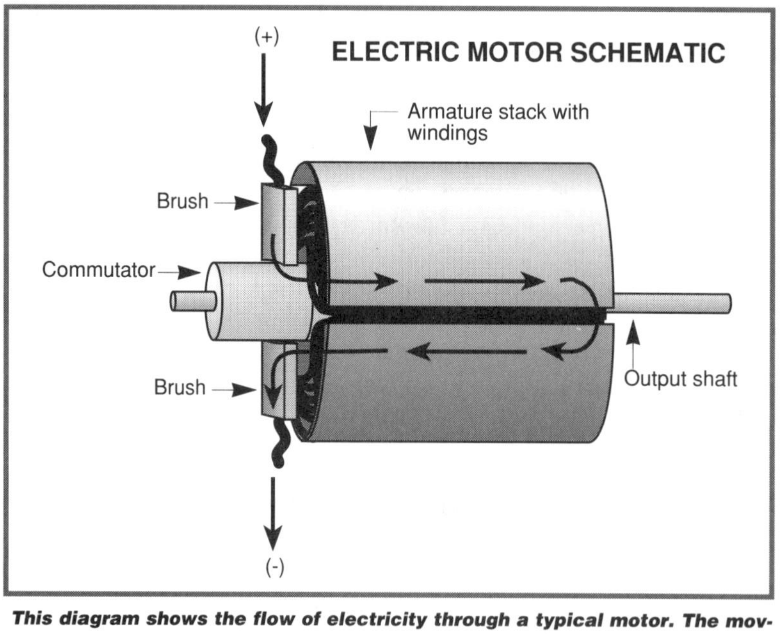

This diagram shows the flow of electricity through a typical motor. The movable contact—brushes to commutator—is the cause of most motor wear and tear; the brushes and commutator should be inspected regularly and repaired or replaced as needed.

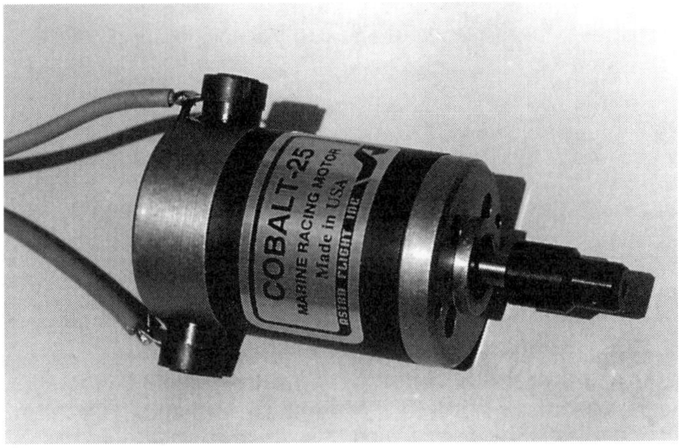

The Astro Flight 25 cobalt motor is popular in direct-drive boats that use 12 cells. This large, 4-turn motor produces plenty of power, and in the correct hull, it can rival speeds attained by certain 3.5cc glow-powered boats.

and because of the nature of fast electric boats, a "heavy" to "extra-heavy" spring is recommended. The strong spring forces the brush more tightly against the commutator, and this increases torque and reduces arcing. Again, replace the factory spring with a heavy one. It's often said that light springs will allow the motor to reach slightly higher rpm, but my own dyno testing shows that this is not usually the case: the heavier the spring, the higher the loaded rpm. Under the high amp loads we use with fast electric boats, light springs almost guarantee arcing and commutator damage.

MOTOR TYPES

As well as being classified according to the composition of their magnets, motors are commonly divided into stock and modified types.

• **Stock motors** are built to meet a certain R/C car performance level by limiting specific design features. The most common motor of this type is the ROAR '91 stock motor—a 27-turn, single wind with bushings, non-adjustable 24-degree motor timing and machine winding to keep the cost down. Other stock motors may have additional timing or fewer turns, but they aren't allowed at car races run under ROAR '91 rules.

• **Modified motors** can have any number of turns, and their design allows their timing to be altered so that their power levels may be fine-tuned. They have ball bearings to reduce drag, and the most expensive ones are hand-wound to ensure consistently even wire placement and armature balance.

The three national model boating organizations group motors into stock and modified classes. For racing purposes, a stock motor is built on the typical "05" car can with bushings and non-adjustable timing. You can use ROAR stock motors, but the fastest boats usually run with an "outlaw" stock type. These motors may have fewer turns or more timing than the ROAR motors, but they're built on the stock car can using bronze bushings. (For more information on this, see Chapter 12: Advanced Tuning Tips.) As this is being written, the APBA is considering limiting all 6-cell stock motors to those meeting ROAR '91 specs—beginning in 1996.

A modified or "open" motor is any motor that does not meet the "stock" definition. In addition to modified R/C car motors, cobalt types and large-can motors may be used. One very popular motor among electric racers is Astro Flight's no. 306 cobalt marine racing motor. For use with seven or eight cells, this motor produces plenty of power and reasonably long run times. Other racers have been successful in the same cell classes with modified car motors that have 12 to 15 turns.

I've had great results with both wet magnet and cobalt motors. If forced to compare the power output of the two types, I'd guess that the four-turn Astro Flight racing motor's power output is roughly comparable to that of a premium, 12- to 14-turn, wet-magnet motor. In general, the cobalt motor produces more torque and will spin a slightly larger prop than a wet-magnet car motor will. On the other hand, if propped correctly, the 12-turn motor will produce higher rpm. Which will be faster? That will depend on the hull type and weight and the skill of the builder, because both types of motor hold national speed records.

The major parts of an 05 R/C car motor. This one features an aluminum endbell and can cap, but in other respects, it's identical to thousands of other motors being run worldwide.

Of course, speed doesn't win races—finishing first does. Either type of motor could be set up to be super-fast for a short time, but when set up in that way, they may not finish a heat before they burn up!

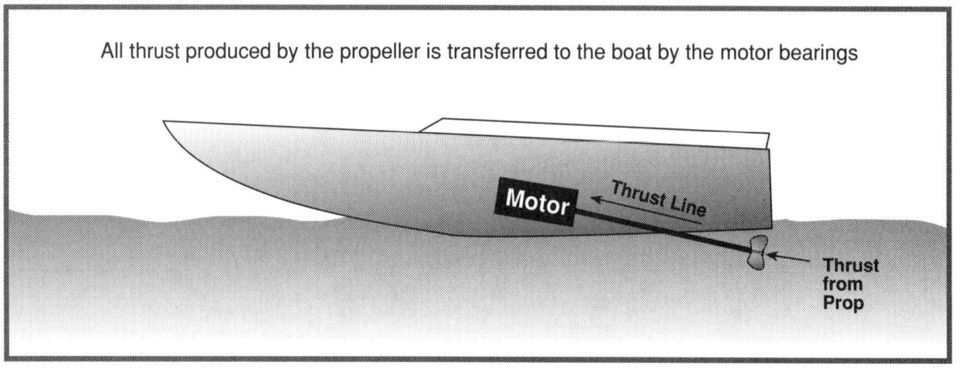

On direct-drive boats, all the thrust produced by the propeller is transferred to the boat through the motor's rear bearings. Some modelers take this strain off the motor bearing by installing a thrust bearing between the motor and the coupler. (See photo on page 76.)

MOTOR TIMING

Motor timing is a measure of when the brushes supply electricity to the commutator relative to the armature's position in the motor can. A motor with neutral timing will have its brush hoods positioned over the center of the motor magnets to take advantage of their maximum magnetic strength to spin the armature. As the motor spins, the magnetic field in the armature stacks needs time to build up, so the armature is "advanced" to allow time for this field to build. With the brushes slightly ahead of the center of the motor magnets, the armature's field is fully built by the time the brushes are centered in the can magnets. A little timing advance is needed for most motors to reach their maximum power potential.

Modified motors can come from the factory with their timing advance set at from zero to more than 20 degrees. ROAR stock motors are permanently timed with 24 degrees of advance; shouldn't the modified motors be timed with at least 24 degrees, too?

A motor's timing should be set to suit its application. ROAR motors are designed to be used in cars that have gear ratios that can be changed to optimize their performance. Advancing a motor will increase its rpm and power, but too much advance will cause overheating. In model boats, a ROAR motor's 24 degrees of timing is probably a little too much for the rpm at which it runs; it will cause excess heat buildup and a loss of efficiency.

The optimum timing for low-turn modified motors still depends on the rpm produced and the current drawn, but between 5 and 10 degrees is a good place to start. Just remember that more advance often means more heat, more power and more rpm but shorter run times. If you use more than 20 degrees of advance, try not to run the motor at part throttle, because that can cause excessive heat and brush wear. I hardly ever use more than 10 degrees of advance (about 1/8 inch) on any of my motors. Using more will seldom help you, but increased heat, reduced torque and brush arcing (radio interference!) might hurt you. Don't run less than zero degrees of timing though, because that can also cause overheating and will reduce power significantly.

Because of their different construction and design, the cobalt motors by Astro Flight have different timing requirements from the more common car motors. Bob Boucher's book, "Electric Motor Handbook" is excellent on electric motors in general and on Astro Flight's cobalt motors in particular. He gives detailed instructions on timing Astro motors, and I encourage everyone who uses an Astro motor to read this book.

BREAK-IN AND MAINTENANCE

There's a great deal of confusion about the proper way to break in a new motor. One thing is certain: to ensure maximum life and power, break-in is vital. The bearings and brushes must be given a chance to wear in without overheating. If you put a brand-new motor in a boat, run it at full speed until the pack dumps then take it apart, you'll probably find the bushings/armature

This armature has the spacer shims installed on its shaft. They're needed to position the armature correctly, without binding, inside the can's magnetic field.

shaft galled and the commutator and brushes burned and damaged. This happened because the motor was forced to run at high amp draw before its parts had a chance to seat properly.

Some car racers suggest that a new motor be broken in for an hour or more to seat the brushes properly, but I think this just puts excessive wear on the bearings and commutator (comm). Most motor manufacturers recommend a far shorter break-in time.

My break-in method is typical of most manufacturers' methods: I run the new motor on four cells for 5 to 8 minutes, stopping halfway through if the motor seems to be getting too hot. Of course, I'm always careful to oil the new bushings or bearings before break-in, and I lubricate the brushes with comm drops, too. This method doesn't fully seat the brushes, but it breaks them in enough to prevent the comm from being damaged when the motor is subjected to full throttle in a boat. The first high-power run will finish seating the brushes.

Break-in isn't just a one-time procedure. When new brushes are installed, they must be broken in, too. The same advice applies each time the comm is turned; use new brushes and break them in. You don't *have* to do this every time, but performance will suffer if you don't.

Motor maintenance is a necessary part of high-performance fast electric boating. Just having a hot motor and batteries doesn't ensure winning performance; keeping the motor in top shape will maximize its power output. At the high amp draws we use, brushes and commutators wear quickly. Buy a comm stick, and use it after every few runs to keep the commutator and brushes clean. Check the brushes often, and never let them wear down by as much as a third of their length before replacing them.

All motors need oil, but too much can be harmful. On bushing motors, use an oil that's specifically for bushings, and put a drop or two on the bushing metal before every run. Don't put too much on the brush endbell bushing, because oil may seep onto the brushes and cause them to burn. Motors with ball bearings require a lighter oil, and one drop on each bearing before every run is plenty.

At the end of every race day, pull out the brushes and use a comm stick to polish the comm's surface and the brush faces. A comm stick is an abrasive rubber tool that can be inserted into the brush hoods and pressed gently against the comm. Slowly turning the armature shaft will abrade a little carbon off the comm and present a fresh metal surface to the brushes. Don't overdo this; a little polishing will be enough. Use the other, rounded end of the comm stick to take the glaze off the brush faces.

After a time, a motor's commutator will have worn enough to require that it be re-cut. Remove the brushes and carefully check the comm's surface for pits, deep grooves, or darkened areas. If you see these, and a comm stick doesn't clean them, re-cutting will probably be required to restore the motor to like-new performance. This is easiest to do on a modified motor, because the armature can be removed from the can and turned on a lathe. Stock motors are usually thrown away when the comm is worn, but they can be re-cut on a stock-motor comm lathe.

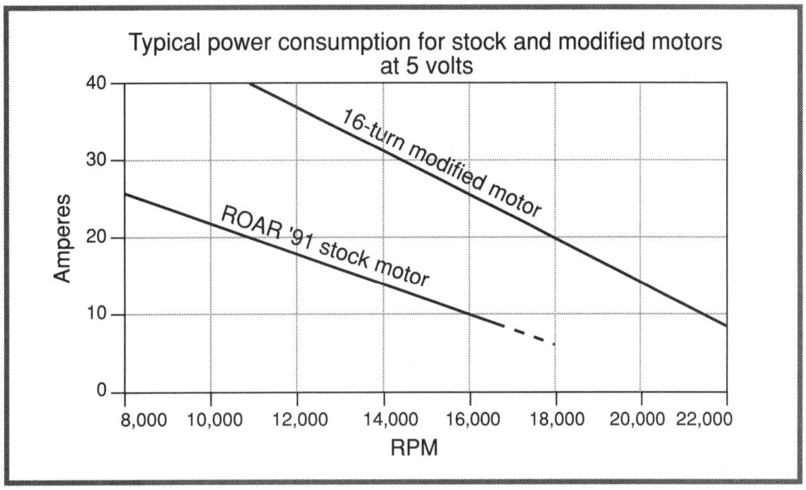

The power consumed by electric motors varies with their design and speed. The modified motor consumes much more power at higher rpm than the ROAR '91 motor does; its power output is likewise much higher. Note that as rpm decrease, amp draw increases. Slow that motor down too much and the high amp draw will overheat it and cause a meltdown.

CLEANING A MODIFIED MOTOR

If you want to work on your own modified motor, you'll have to take it apart to clean it fully and check for wear.

Follow these steps:

- Remove the brush springs and pull the brushes out of their hoods.
- Carefully mark the can and endbell with a felt-tip pen so you'll be able to reassemble them correctly later!
- Loosen the endbell screws by about three turns, and rotate the endbell until it can be pulled off the can.
- Pull the armature out, being careful to note the number and position of the small spacers at each end.
- After wiping out as much brush dust as you can, flush out the can and endbell with motor spray.
- Check the bearings and comm, then reassemble the motor with the correct number of spacers on the armature.
- Turn the endbell until the marks you made line up, then tighten the endbell screws gently—too tight, and

This mono catches some air as it runs down the front straight.

you'll cause damage.

- Replace or reassemble the brushes and oil the bearings.

For model boaters, corrosion is always a danger. The parts most susceptible to rusting are the ball bearings, and even frequent oiling may not prevent them from being damaged. In a high-power motor, a small speck of rust in a ball bearing can act like a grain of sand and quickly destroy the bearing. After a day at the pond, check the bearings by removing the armature from the can and spinning each one with the point of a pencil. A bad bearing will feel rough or sound noisy; for top performance, replace it.

All this applies to the usual wet-magnet car motors. If you run a cobalt motor (such as those made by Astro Flight), your maintenance schedule will be very different. Owing to the way in which they're constructed, these motors last a long time with minimum maintenance. Of course, their bearings need oil just as a wet-magnet motor's do, but the special Astro silver brushes last a long time; at 40 amps, Astro claims that they will last 100 hours! Though I can't verify that, I know from experience that they will run for far, far longer than any wet-magnet motor brush will run before it has to be replaced. The Astro commutator is also thicker and larger in diameter, and it seems to become burned and scored only if badly abused. Though these motors are more expensive, they save you time and the cost of replacement parts.

MOTOR RECOMMENDATIONS

I hesitate to recommend specific motors for fast electric boats because so much depends on the type and weight of a model, how well it has been set up, and the experience of the modeler. But "Which motor?" is one of the most frequently asked questions about fast electric boats, and avoiding the topic would be a cop-out. I'll qualify my advice by saying that, in my experience and that of many other boaters, the recommendations I give are good ones, but there's no guarantee that the motors recommended will work well in your application. Any of these motors can be made to work poorly by using the wrong propeller or by attaching it to an inefficient drive system. With that out of the way, here goes:

- **Direct-drive monoplanes** with submerged propellers can use anything from a 27-turn ROAR motor to perhaps a 16-turn modified motor. If more than eight cells are used, a smaller prop or a milder 18-turn motor is recommended to keep the amp draw down. Light, surface-drive monos with six or seven cells can use a 13- to 15-turn motor. Upping the cell count to eight makes a milder, 14- to 16-turn motor more attractive.

- **Boats with gear drives** can use similar motors, but if 12 cells are used, it's advisable to have at least 15 turns, except with premium hand-wound motors; at this cell count, the extremely high armature speeds (over 50,000 rpm) experienced with motors that have fewer turns can destroy bushings or low-grade bearings.

As the voltage increases, so does motor rpm, so use a seemingly milder motor. Since a low-turn motor spins faster at a low voltage than a milder one does, it follows that, at higher voltage, the low-turn motor will run even faster. For direct drive, motor speeds of more than approximately 25,000rpm usually aren't too practical owing to the smallness of our prop blades, and when running a motor with more turns at higher voltage, the rpm produced are kept within reasonable limits. A 13-turn motor that may produce its maximum efficiency at 22,000rpm on six cells could spin up to 30,000rpm on eight cells—way too high for direct drive. Substituting a 15-turn motor reduces rpm, but overall power is still increased because the voltage is higher, and the motor spins the prop faster than it did on six cells.

COUNTING CELLS

A note about cell count: single 05 motors may not run well on more than eight cells, especially in direct-drive. If you want to use 12 cells with a car motor, a twin gear drive wired in parallel is probably best; it divides the amps evenly between the motors, allowing each one to produce its high rpm at 14 volts while drawing only half the amps. The net power is about the same—without abusing one motor on 12 cells pulling maximum amps.

Astro Flight offers a 12-cell motor that produces plenty of power, and this is another popular option for 12 cells. The power outputs of the Astro Flight and a dual gear drive with high-quality motors are very similar. A low-cost alternative to either is the Graupner Turbo 700 Series motor. This inexpensive motor runs well on 12 cells and produces a surprising amount of power, but it isn't as expensive or complicated as a gear drive or a cobalt motor. The Graupner's power output is not of the same level as that of the Astro or gear drive, but if cost is a major consideration, the Turbo 700 motors can be attractive choices.

Hydros can use motors with slightly fewer turns than monos can, but a motor with fewer than 12 turns is probably a poor idea, especially in direct drive. Motors with nine to 11 turns must spin at too many rpm to be efficient with the small props we use; spin a 27mm-diameter prop at 30,000rpm, and its tiny blades won't grab the water well. You could use a larger prop, and some racers do this, but they may have to rebuild their motor after one run because of the excessive number of amps drawn at lower rpm.

Geared hydros running six cells can effectively use 10- to 12-turn motors, because at this lower voltage, the gearing allows a larger, more efficient prop to be used. Still, the speed advantage over a 13- to 15-turn motor will be small, and maintenance requirements will be greater with the lower-turn motors. If you don't race at national events, there's little practical use for motors with fewer than 13 to 15 turns.

Choosing an Astro Flight cobalt motor is much easier; Astro recommends that you choose according to the number of cells you use. Though it's possible to use more cells than recommended, with the newer 4-turn Astros, this isn't wise. Astro states that the no. 306 8-cell motor does not run well on 12 cells because with that number of cells, it must run at excessively high rpm. Follow the maker's suggestions and keep out of trouble.

MORE MIGHTY MOTORS?

When you've gained experience with your model, you may want to experiment with more powerful motors than those already suggested. You might find that performance can be improved slightly, and that's good. On the other hand, you may discover that a lower-turn motor gives too

short a run time, or that it requires major maintenance after every day's running. You may go a bit faster, but

the costs both in dollars and inconvenience may be too high to pay.

One motor recommendation is less controversial. If you want to run your model fast, but you aren't interested in racing, buy a good-quality, machine-wound modified motor to replace the stock type found in most kits. A 20-turn mod motor will last longer than the stocker and will produce more power. Well-oiled bearings won't wear out as quickly as bushings will, and by running with less timing advance than a ROAR '91 motor, the mod motor will run cooler and have better low-end punch. If you want more speed for fun running, select a motor with fewer turns. The less expensive mod motors are excellent choices for fast electric boaters who want to upgrade without spending a fortune.

For national racing, or for the highest speeds, the increased cost of the hand-wound modified motor is justified. The motor builder spends more time on the hand-wound motor; the windings are tighter and more even; the armature is balanced for less vibration and more power; and the magnets, bearings, brushes and springs are better. This extra quality costs money, but the result is a motor that will simply produce more power than the machine-wound types. These premium motors demand more maintenance than their less expensive cousins, but that's one of the prices we must pay to go faster. The hand-wound motors are not necessary to go fast, but for that last bit of speed and power, they can be worth their price.

CHAPTER 4

Batteries

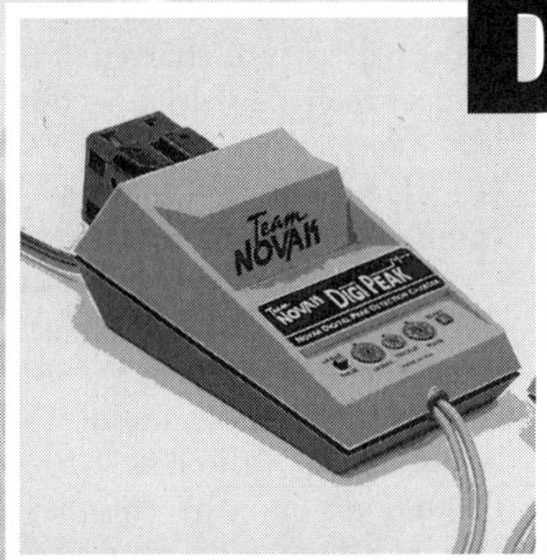

4 BATTERIES

BASIC THEORY

If you ask an average model boater what powers his boat, the chances are pretty good he'll say that it's his motor. Actually, that's the wrong answer: all the energy an electric boat possesses is in its battery; the motor simply converts that energy into work.

Fast electric boats almost universally use nickel-cadmium (Ni-Cd) batteries for power. Though other battery types are being field tested, it's a sure bet that Ni-Cds will continue to be our main power source for some years to come. They store lots of energy, can be recharged quickly and are not too expensive. Actually, without the Ni-Cd battery or an equivalent, there would be no fast electric boats as we know them. Do not underestimate the importance of the battery to your boat's performance!

The most important feature of Ni-Cds for fast electric boat use is their low internal resistance, which means that they can deliver their energy rapidly. Older battery types stored as much or more energy, but owing to their design, they could only be discharged at a much slower rate. Slow energy delivery means lower horsepower for the motor at a given voltage. That's one reason why electric records set 20 years ago with motorcycle batteries needed high voltages (36 volts or more) just to equal current records set with 7.2 volts. The other main reason for the much better Ni-Cd-powered model performance is the greater weight of the wet cells used in the older boats.

To increase cell performance and enhance their ability to take a charge, batteries should be fully discharged immediately after each run. Many modelers use a battery dumper made with automobile light bulbs. This small unit uses six GE 1157 bulbs for a discharge rate of about 12 amps. Some dischargers use as many as 12 bulbs, but care must be used, because the high discharge rate may overheat a battery that's still warm after running.

Ni-Cds can also be recharged quickly. The old, lead-acid batteries I used when I started in the hobby 30 years ago took an entire night to charge, and because of their high internal resistance, they discharged their power too slowly to develop much speed, especially with the primitive electric motors of that time. Because of their lower internal resistance, the Ni-Cds we use today can be recharged in less than an hour, and they discharge quickly. That's why we can go fast many times a day instead of waiting overnight for a charge.

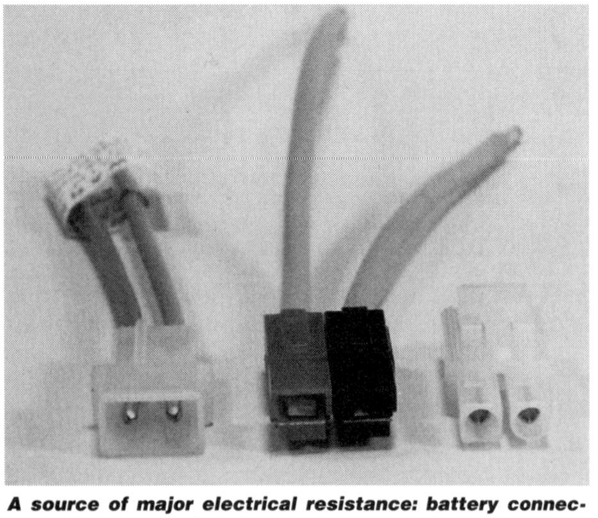

A source of major electrical resistance: battery connectors—especially those that come attached to most packs such as those on the ends in this photograph. The center connector is the high-performance model (available from Sermos, Lite Speed and others). Its resistance is almost as low as an equal length of wire. The "car" connectors on the ends can get so hot during marine use that they will melt together. Replace them.

Most Ni-Cd batteries produce about 1.2 volts per cell. A 6-cell battery pack is rated at 7.2 volts, or six cells times 1.2 volts. Obviously, the more cells there are in a pack, the greater the voltage supplied, but weight increases as well. Cell voltage drops as the battery is discharged; it starts at more than 1.2 volts and ends at less than 1.1 volts, just before the cell "dumps—is out of energy—completely (near zero volts).

The most common battery-cell size is the sub-C. Battery capacity is rated in amp hours, and the usual cell has a capacity of around 1.4 amp hours (usually written as 1400 milliamp hours or mAh). This means that, at a discharge rate of 1.4 amps, the battery will dump in one hour; at a rate of 14 amps, the battery will last 6 minutes, etc. Actual run times will be slightly shorter owing to differences in efficiency, but for our purposes, these figures are close enough.

The maximum power available for your boat is based on a battery's voltage (number of cells) and the current draw (in amps). Most kit boats are set up for a current consumption of 10 to 20 amps, which gives run times of 4 to 8 minutes. This current use won't overheat the average kit motor or speed control, and it's generally trouble-free for the beginner. Of course, speeds at this discharge rate aren't too fast; if we want to go truly fast, we need more electrical power, and that means more cells or a higher amp draw.

If we raise the current to between 30 and 40 amps, speeds can increase substantially, but there are drawbacks: shorter run times, higher temperatures and increased wear and tear on our motor and speed control. Keeping the current draw the same and increasing the voltage by adding cells will improve performance and potentially subject our equipment to less wear and tear.

Two 8-cell monos round a turn.

ELECTRICAL POWER

The relationship between volts, current and watts.

Voltage: volts
Current: amps
Power: watts or horsepower
Volts x amps = watts
Watts ÷ 746 = horsepower

Examples:
7.2 volts x 20 amps = 144 watts
7.2 volts x 40 amps = 288 watts
14.4 volts x 20 amps = 288 watts

Note: these are theoretical figures that do not take into account a system's resistance losses.

BATTERY TYPES

The three most common types of Ni-Cds are the SC, SCR and SCRC cells.

• **SCs** are sport batteries, and they're usually the kind supplied with basic kits. They're inexpensive and durable, and for the beginner, they work well enough. Designed originally for industrial use, they have a higher internal resistance than some other types. Because of this, they will not produce as much power for a motor, and they are more apt to overheat when charging and discharging.

• **SCR cells** are the next step up; they have about half the internal resistance of the SCs, so they discharge much faster, and that makes more energy available to the motor over a shorter time. This means that the motor produces more horsepower and the boat goes faster. The red Sanyo* SCR is more costly than an SC, but to the fast electric enthusiast, its performance makes it worth it. An SCR can be recharged many times a day, so with two or three packs, you can run for many hours.

This battery fact was brought home to me several years ago when I was running a mono I had built for my wife. She had been using SC cells, so that's what I used when I took the model to the pond to try out a new motor. After running out two of her packs, I wanted to do more propeller testing, but I didn't have any more of her batteries. I did have one of my SCR racing packs along, so I plugged that in and launched the model. What a difference! The boat was much faster than before, and the only difference was the SCR battery.

The SCRs are currently the choice of most serious fast electric racers, and they're highly recommended if you want to go fast. It makes little sense to buy the most expensive motor and speed control, then handicap yourself with substandard batteries. The SCs work OK, but for serious racing, use the SCR cells.

• **SCRCs** are the newest type of cell. They're the same size as the SCRs, but they have a higher capacity—usually 1700mAh. This means they'll run longer and produce a high rate of discharge, but they should be limited to only two or three recharges per day. Their internal resistance is

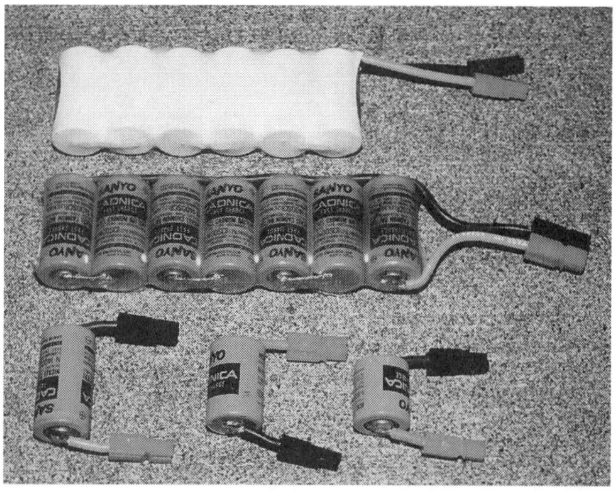

This sample of the batteries offered by Electric R/C* includes the three most commonly seen cell sizes. Front of photo, left to right: the standard 1400mAh sub-C cell, 1000mAh and 600mAh cells. The two smaller cells are of limited use for most fast electric boating, but they can be useful in some circumstances. In four-lap heats, the lower weight of the 1000mAh cells can enhance speed, if sufficient run time can be obtained. The tiny 600mAh cells are most often used for drag racing. Note the heavy battery bars used to connect the cells in the assembled pack. This reduces battery resistance (compared with the small welded tabs), and that means a higher voltage is available from the cells.

also apparently slightly higher than that of the SCRs. If you prop your boat to discharge an SCRC pack in less than 2 minutes, you're drawing an average of more than 50 amps and are probably abusing your motor and speed control. High-capacity batteries are great for fast sport boating and endurance events, but be careful if you use them for racing, and watch for signs of overheating.

Some top racers prefer special matched and pushed cells, such as those shown here, but average fast electric boaters don't usually need them.

MATCHED AND PUSHED PACKS

Matched and pushed packs are popular with car racers, some of whom believe that they're needed to win.

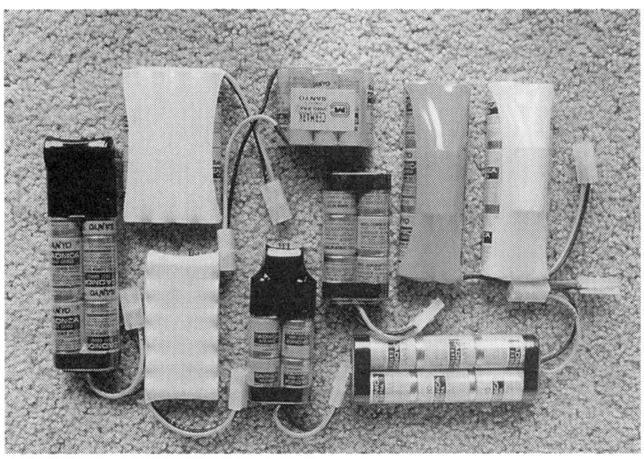

As shown here, battery cells can be assembled into packs of a variety of sizes and shapes.

They're more important for top-level car racing, but their advantage is lost for most fast electric boats. Most boat heats last less than 2 minutes, so the extra capacity—usually only 15 to 20 seconds at 20 amps (the more you pay, the more you get)—is of little importance. Only if your boat is propped to pull 40 amps or more should the extra capacity of the matched and pushed cells be an advantage. They won't hurt, of course, but I don't think that they're really needed for most fast electric boating.

An additional fact: a pack that was matched when new won't be matched several dozen recharges later. Individual cell capacity will drift somewhat over the pack's life, so last year's pack won't match a new pack this year. For most use, I recommend that you buy the less expensive packs; the money you save could be used to buy an extra pack to allow you to spend more time testing or practicing.

I can't argue with the fact that a great many speed and racing records have been set with special packs. Some have also been set with randomly selected bulk batteries. If you're going after a speed record, or to a national race, the little advantage offered by an expensive pushed/matched pack may make the difference. Just don't feel that you have to spend a lot of money on special batteries just to run fast.

The biggest advantage matched packs offer is one of quality control. Batteries are made by the tens of thousands, and a poor cell gets out occasionally. Most bulk-type packs are assembled with random cells, and it's possible to get a pack with a bad one. In matched packs, each cell has been tested before assembly, so the chances of getting a poor cell are greatly reduced. Of course, most bulk packs don't have bad cells either; the matched packs just ensure that you won't get one.

By far the most popular battery size for R/C models is the sub-C cell. For most fast electric boating, this is the best battery to choose, because it

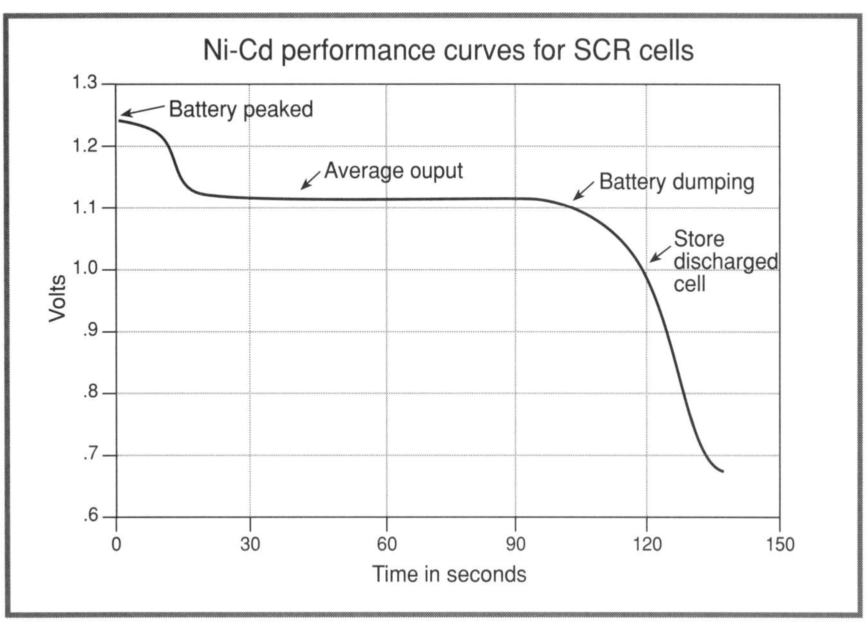

Left: if properly charged, some Ni-Cd cells will deliver almost peak power for most of a run. The preferred SCR cells don't lose much output until the end of a run, and then they dump rapidly. A sudden, almost complete drop in power is a mark of a good pack; it signifies that its cells are well-matched.

provides good power for a reasonable time. In some specialized instances, though, you may not need the high capacity of 1700mAh or even 1400mAh. For drag racing, in which run times are less than 10 seconds, the sub-C cells are needlessly heavy. Also, in some oval racing, a fast model running sub-Cs may be able to finish a heat with plenty of battery charge left. In these instances, a smaller, lighter battery may be an advantage.

The chart below shows the specifications of various types of SC cells, from the tiny 600mAh through the 1700 sub-Cs. For four-lap IMPBA heats, the 1000mAh cells usually last long enough if the boat has been set up extremely well; but for five-lap heats, they're marginal. By using 1000mAh cells and saving several ounces in an already very light boat, the power-to-weight ratio can be increased quite a bit. For drag racing, 600mAh cells have plenty of capacity and save even more weight—always the enemy of any dragster. Though these smaller cells do have a slightly higher internal resistance than 1400s, for most boaters, the weight reduction more than makes up for the slight reduction in delivered voltage.

BATTERY SPECIFICATIONS

Type	Weight (oz.)	Minutes at 30 amps
600	1.1	1:12
1000	1.4	2:00
1400	1.8	2:48
1700	1.9	3:24

Note: using the last 15 to 20 seconds of battery power can cause the battery to overheat. Repeated overheating will damage the cells, and with time, they'll lose capacity.

CHARGING AND MAINTENANCE

The recommended procedure for charging Ni-Cds seems to change yearly, but as this is being written, this is most common:

The batteries are charged at 4 to 5 amps until their voltage peaks, or until they start to get warm. Charging at a higher rate, 9 amps for instance, is supposed to give a higher discharge rate, but most experts feel that this is not true, and that the high charge rate can damage the battery.

If a battery is charged at 5 amps, the cells should be fully charged in about 17 minutes. If you know that the pack is flat (less than 1.1 volts per cell), you can use a timed charger set at 17 minutes to charge the pack, but this can be dangerous. If the battery was *not* completely dead, it might be overcharged and get very hot. You can make batteries vent by overcharging them (I know because I've done it!), so I recommend chargers that don't rely solely on a timer. Still, if you're careful, a timed charger can work well. A tip: charge a cell when it's cold; when it starts to get warm, it's fully charged.

Advanced chargers, such as these by Novak, reduce the likelihood of damaging a battery by overcharging, but they are not essential for top performance. When buying a charger, be careful to check the maximum number of cells the unit will charge at one time. Many car chargers will not charge more than seven cells.

From a wall outlet, this Astro Flight charger will charge up to eight cells; when connected to a car battery, it will charge more than 12. Peak chargers like this allow the modeler to "fire and forget"; just plug in the pack, push the start button and walk away.

• **Which charger should you use?** It depends on how you intend to use your battery, but I've had excellent luck with peak-detector chargers. They have special circuitry that detects when the battery is fully charged; the charger then shuts itself off. With these chargers, you don't have to set a timer or watch a gauge, so it's almost impossible to overcharge your battery. Some peak-detector chargers will "false peak," so it's a good idea to check each pack. After the charger has switched off, wait a few minutes and push the start button again. If the pack was indeed peaked, the charger will shut down in about a minute. If the charger "false peaked" the pack the first time, this repeat run should result in a full charge.

• **Overheating.** Batteries are most often damaged by overheating. If a cell becomes too hot, its internal gas pressure will rise so high that it will "vent" some of its stored fluid and permanently reduce its capacity to take and hold a charge. Heat is generated both by overcharging and by discharging at a high rate. To reduce the risk of damage, a battery should be cool before it's charged and cool again before it's run. I will often put a hot pack into a sealed plastic bag and place it in a pail of water before charging it.

• **Battery "memory."** This is often discussed; it simply means that a battery "learns" how to take a charge and how to discharge. (Of course, the battery has no intelligence and can't actually learn, but the analogy is pretty good anyway.) If you charge a pack then run it down only halfway, recharge it and run it down halfway, etc., after a time, it will discharge halfway and then dump, so effectively losing half of its rated capacity. The cell's memory can be changed by fully charging the battery, fully discharging it, then repeating this cycle several times.

For top battery performance, you should fully discharge each pack before you recharge it. This battery "dumping" ensures that the cells do not pick up any unwanted memory effects. Some top-end chargers have discharge circuits, but many racers use car stoplight bulbs to drain their batteries before recharging them. Remember that the battery will heat up when you discharge it, so be alert to overheating. After a run, you may need to let the pack cool off before you dump it.

You can improve the performance of your packs by "balancing" them regularly. To balance a pack, charge it fully as usual, then put it on a trickle-charger overnight. This procedure corrects for any cell that might take a charge more slowly than the rest of the pack and gets it up to full charge without overheating the rest of the pack. After being balanced, a pack will have a higher overall capacity; it won't start to reduce its output a minute before it dumps totally. To check how well your packs are balanced, run them to discharge, carefully watching how your model is running. If the boat runs at full speed until it suddenly dumps, you know that your pack is well-balanced and will produce maximum power for as long as possible.

If you won't be using your batteries for more than a few weeks, it's best to store them in a discharged, but not totally flat, state. The experts don't agree on the exact level of discharge you should aim for, but I think that if stored while still charged at about 1 volt (1.25 volts is normal for a fully charged cell), the cells will suffer only minimum damage from extended storage.

I can't tell you how long your packs will last, because so much depends on how you use or abuse them. With proper use, an average pack should last for years of sport running—certainly many hundreds of charge/discharge cycles. Battery manufacturers put the life of most Ni-Cds at between 500 and 1,000 charge/discharge cycles. For top-level racing, premium pushed and matched packs will drift off their original values with time, but they'll still deliver plenty of performance for quite a while.

If, on the other hand, you overheat your packs as a matter of course, they'll soon fail to perform well enough in tough competition. It's best to sell them cheaply to other boaters with lower performance expectations and then buy some new packs. This time, be more careful with them and they should last much longer.

CONNECTORS

Even the best batteries won't deliver the performance you paid for if you try to route their power through cheap battery connectors. The standard Kyosho and Tamiya-type connectors found on most packs just aren't good enough for high-performance use. At the high-amp current draws used in fast electric boats, poor connectors will arc and get so hot that they may actually melt together. A hot connector is one that's causing excessive resistance, and that means that your boat's motor is not getting the full battery voltage. Each Tamiya-type connector can reduce the power going to your motor by ½ volt.

Though there are quite a number of available connectors, the ones I know best are by Sermos*, Lite Speed* and Astro Flight. These feature plated connectors (silver or gold) and firmly held contacts. Other connectors may work well, too, and new ones are introduced yearly, but I know that the ones by these three companies will work well for any fast electric boater.

CHAPTER 5

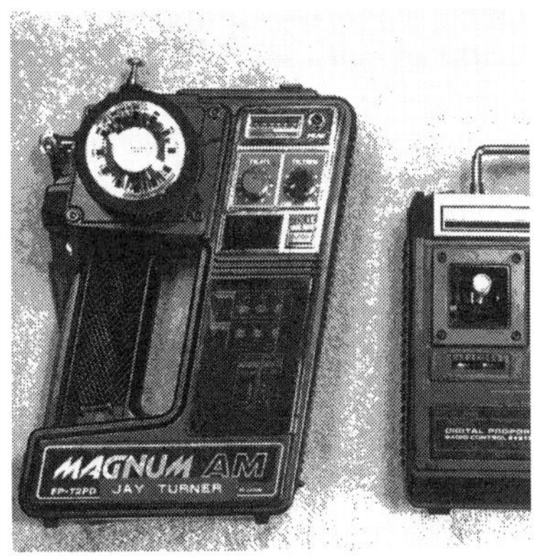

The Radio System

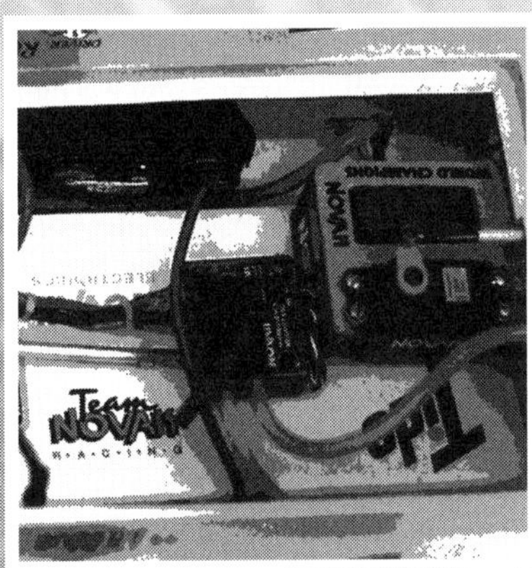

FREQUENCIES

R/C boats are required by the Federal Communications Commission to run on certain frequencies. As this is being written, surface vehicles (cars and boats) must use either the 27MHz or the 75MHz band; the 72MHz band is reserved for model aircraft only, and the 27MHz band is shared among all R/C models.

It's very important not to use an aircraft frequency for your boat. If an R/C airplane is flying in your vicinity (perhaps unseen by you) and you turn on your transmitter, if the frequencies match, the aircraft could crash. Though a model boat that runs up on shore owing to frequency interference is usually repairable, an aircraft that "augers" in from 100 feet probably isn't. Don't risk another's model or personal safety by violating the law; use a surface frequency when you run a fast electric boat.

TRANSMITTERS

Today's radio systems are a far cry from what was available 20 or more years ago. Back in the '60s and early '70s, most affordable radios were rather rudimentary, often having a single transmitter button for sequential rudder-only control and a complex switching device at the receiver to provide throttle. The leading technology was found in upscale radios for R/C aircraft and, reflecting that background, the transmitters had stick controls that resembled those used in full-size aircraft. As sales and competition increased, prices decreased, but most transmitters were still tailored to suit airplane enthusiasts.

In the mid- to late-'80s, as the R/C car hobby bloomed and manufacturers enjoyed massive sales, radio technology advanced dramatically. This benefited everyone in the R/C hobby, because the increased sales allowed the manufacturers to reduce the unit cost of each radio, and that made them even more popular. R/C boat hobbyists quickly saw the advantages of the newer car-type radio systems; cars and boats need only steering and throttle control—not the four or five controls that aircraft need.

Today, transmitters come in two basic styles: stick and pistol-grip. Stick transmitters are a throwback to the aircraft roots of the hobby, but they remain popular with a small segment of fast electric boaters. Most boaters favor wheel-type radios because they're more realistic; full-size boats have steering wheels, and many racers feel that R/C boats should, too. Most of us find it easier to control a fast boat with the wheel controller, and unless you're firmly wedded to a stick transmitter (perhaps because of experience with R/C aircraft), I strongly suggest that you invest in a radio system that has a wheel/pistol-grip transmitter.

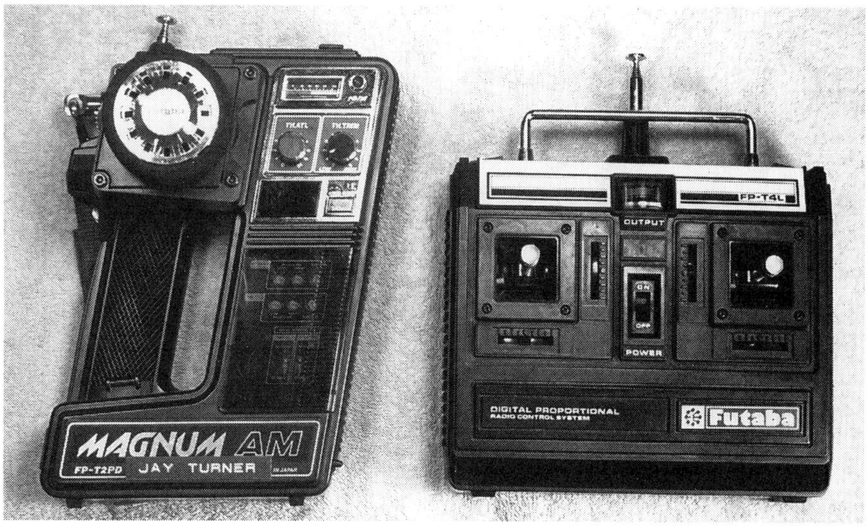

The two types of transmitter used by boaters are the pistol-grip and the stick. Unless the modeler has experience with the stick transmitter, the pistol radio is recommended. The higher-end radio on the left is not necessary for top performance; I used an inexpensive model for years without cause for complaint.

RECEIVERS

The R/C receiver is the link between the boat's controls and your transmitter, and it must function perfectly if you want good control of your model. A potential problem exists when several boats are run together (for example, during a race in which the transmitters are physically close together). This can cause interference for your boat's receiver, and it might respond to controls from both your transmitter and from another driver's radio. The result will be poor control at best and a crash at worst. (The receiver won't be able to decide which signal to respond to, so you partially lose control.)

Radio manufacturers recognized this problem and, in response, introduced narrow-band receivers in 1991. When "listening" for a signal, these are more restrictive, and they're less likely to "hear" another driver's transmitter. The problem of interference hasn't been solved, but the narrow-band radios have helped. Though it was hoped that the new receivers would allow drivers in the same heat to use adjacent channels simultaneously (for example, channels 75 and 76); in practice, the adjacent channels could still cause interference problems. Most clubs don't allow racers to run on either odd-numbered channels or adjacent channels.

CRYSTALS

On the surface, changing frequencies is a simple matter: just pop out the radio's transmitter and receiver crystals, and insert the appropriate new ones. This usually works well, but be advised that if your radio was originally purchased with, for example, channel-74 crystals already installed, it was tuned at the factory for that frequency. When you substitute channel-90 crystals, the radio is not optimally tuned for the new frequency. If you stay reasonably close to the original crystal frequency, however, the radio will still be well-tuned. This is important to know when you buy extra sets of crystals, particularly if you race (you want to be able to change frequencies to race in a particular heat when two or more boats are on the same channel). One or two spare sets of crystals come in handy, but it's still best to choose channels that are close to your radio's original crystals. If you can't do that, you can send the radio back to the manufacturer to be retuned to the new frequency.

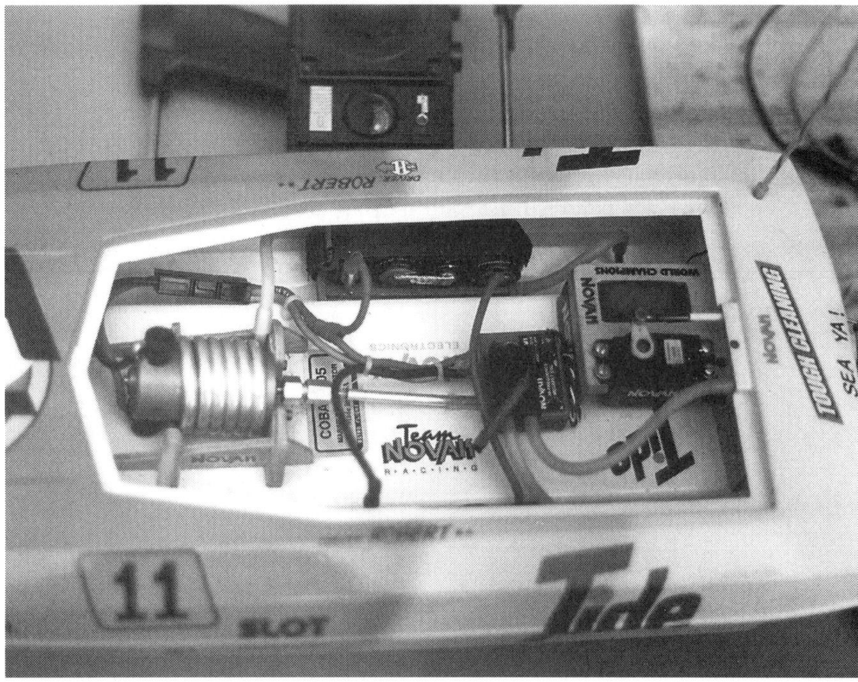

This installation places the receiver out of sight between the servo and the ESC. Although it works well for this modeler, it might be better to move the ESC several inches away from the receiver to reduce the chance of interference.

INTERFERENCE

Even with a correctly tuned radio, modelers will occasionally notice that their boats don't respond well to the controls. Sometimes, this may be the result of powerful radio interference, such as at a pond that's near a paging company's transmitter tower. Little can be done to eliminate this type of interference—other than running at another site.

There are, however, several things you can do to limit other types of radio interference. A common mistake is to cut the receiver antenna short so that it will fit inside the boat. This is a bad idea, and it almost guarantees that you'll have a reduced radio range and be susceptible to interference from other radios as well as from non-R/C sources. If you elect to use the stock receiver antenna, run it through an antenna tube, leaving as little of the wire loose inside the model as possible. If the wire is longer than the antenna tube, let the extra length trail outside the boat; just make sure that it can't foul the model's propeller!

An alternative to using a long external antenna is to wrap some of the wire around a cardboard form, as shown in most radio instructions. This allows a shorter length to protrude out of the model, but it restricts your range somewhat. I suggest that, if you want an external antenna that's shorter than 12 inches, consider using a special loaded antenna such as those offered by Deans*. They require that the receiver wire be cut short and soldered to a special plug. I've used Deans antennas for years with complete satisfaction. I can't stress enough the importance of good antenna installation. Many—if not most—radio-range and interference problems can be traced to poor antennas.

There's much potential for radio

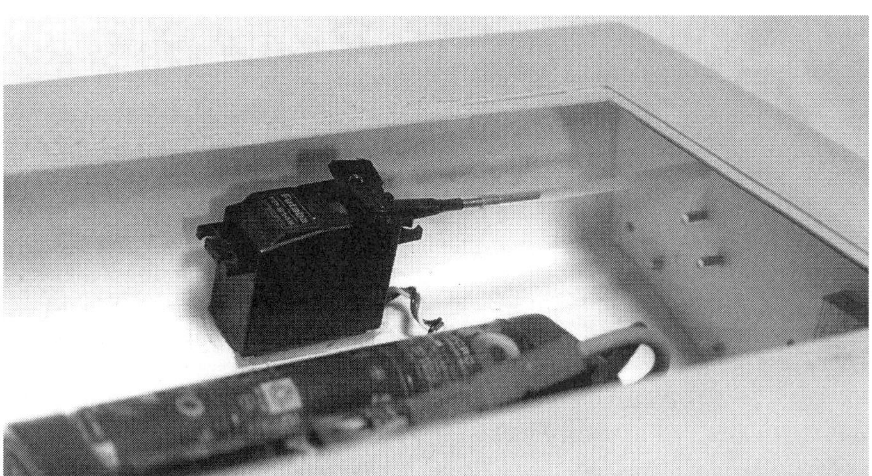

The ultimate in simple servo installation. To reduce binding, make sure that the pushrod follows a straight line from the servo arm to the rudder horn. Silicone glue secures the servo to the hull bottom.

interference within the boat itself. The potent electric motor generates a magnetic field, and the sparking commutator can produce radio interference, too. The wiring that carries the motor's power is also a source of trouble. Of course, you can't get rid of the motor or the wiring, but you can place the receiver as far away from them as possible. In addition, the motor should have at least one capacitor soldered across the terminals to reduce sparking. Some ESCs (electronic speed controls) require the use of two or three capacitors, but even if you use a mechanical speed control, you should use at least one capacitor.

This Graupner Tornado has limited space inside for radio installation, but it can be done.

BECs

Some receivers can use battery-eliminator circuitry (BEC). All receivers require electricity, of course, but they don't always have to be powered by dedicated receiver batteries. Most ESCs are wired to supply the receiver with power. A standard 4-cell receiver pack weighs 4 ounces; that's ¼ pound, and it will slow your boat down (especially troublesome in races). Special, tiny, rechargeable Ni-Cd packs are available, and these will allow the use of a non-BEC receiver; but they aren't cheap, and if they aren't adequately charged before a run, they'll leave you without radio control!

To power the receiver, the BEC uses the electricity in the drive batteries. Usually, an ESC will supply the electricity to the receiver, so the battery socket in the receiver is left empty. One drawback of a BEC system surfaces when you use a very powerful motor and a large prop: the motor's momentary current draw when taking off at full throttle can be so great that the receiver doesn't get enough electricity to operate and you lose control. I've occasionally seen this happen, and it can be very frustrating. The two solutions are either to use separate receiver batteries or to fit a Stutter Stopper from Novak Electronics*. This tiny electronic device stores a little electricity when the battery is hooked up; if the motor draws too much current for normal receiver operation, the Stutter Stopper feeds the necessary electricity to the receiver for a few moments until the motor speeds up and the amp draw is reduced.

DRYING OFF

If your receiver gets wet in a flip, all is not lost. Open the case and remove the circuit boards; pull out the crystal, and shake off or blow off as much water as you can. Use a hair dryer to dry the electronics, but don't get carried away and heat the parts too much. If you can't touch the circuit boards, they're already too hot! Let the parts sit out of the case overnight, then reassemble the receiver in the case and hook it up to the rest of the system. With luck, everything will function well. Next time, put the receiver into a balloon, or seal the deck better. One comment: if you douse your receiver with salt water, all bets are off!

SERVOS

Most fast electric models use standard car and airplane servos for rudder control. Though special, high-speed, quick-reacting servos can be used, they usually aren't needed for oval racing. The most common servos supply around 40 oz.-in. of output torque, which is plenty to turn almost any electric model boat. These servos usually weigh about 1.5 ounces and are very rugged and inexpensive.

Some modelers want to reduce boat weight as much as possible, and this is usually a great idea, because, in boat racing, weight is the enemy of speed. But they can run into problems when they try to substitute a smaller, lighter servo for the standard type. A special, lightweight servo for a small car may save nearly an ounce compared with a standard type. Unfortunately, those tiny servos don't have the power of the standard type; often, their output is only 25 oz.-in. or less.

The problems develop when this small servo is used in a very fast model with BEC. First, the high speeds of a 7- or 8-cell hydro put lots of resistance on a turning rudder, and the servo may have a hard time making the hydro turn. Second, unlike a car, a fast electric boat often remains under full throttle as it enters a turn. The rudder starts to turn, the boat slows, the motor's current draw increases and, suddenly, the electric power supplied to the servo is dramatically reduced. The rudder straightens out or glitches rapidly, and the boat plows ahead without rudder control. I saw this happen to a club member's extremely fast hydro during his first race with the new little servo, and his boat was destroyed when it hit the bank at nearly top speed. Save the trick servos for lower powered models or for aircraft.

CHAPTER 6

Speed Controls

6 SPEED CONTROLS

VITAL EQUIPMENT

In all but the simplest boats, some method of controlling a model's speed is necessary. During a race, a boat may have to be throttled down for the turns or to avoid hitting another boat. Even when you're sport running alone, being able to moderate your boat's speed makes it more versatile and therefore, more interesting.

This 12-cell hydro has no variable speed control—just an on/off switch. One disadvantage of this simple setup is obvious in the photo: full speed makes the prop cavitate excessively, and acceleration away from shore is very slow.

Electric motors can be best slowed down by reducing the voltage supplied to them. The simplest way to do this is with a mechanical speed control; a command to reduce throttle causes a servo to move a set of contacts and place a single resistor in the power circuit from the battery to the motor. Depending on the resistor's value, speed can be reduced slightly or a lot. Another type of mechanical control uses a wiper arm traveling across a wire-wound variable resistor in the power circuit. This supplies variable speed from stop to full. Other mechanical controls switch between parallel and series circuits when twin battery packs are used.

When our local club began to race on an indoor swimming pool in the winter, we soon learned that simple mechanical controls quickly burn up. It was not unusual to see one or two boats stalled in the middle of the pool, quietly smoking as the controller and the wiring smoldered. The reason for the high mortality rate was the tight course we ran; it required a lot of throttle manipulation. The solution was to replace the mechanical controllers with electronic speed controls (ESCs).

The problem with most mechanical controls is that they reduce the power going to the motor by turning some of the battery's electrical energy into heat. This makes them inefficient, because the wasted electrical energy would be better used driving the motor. All that heat has to go somewhere, and it often stays in the controller until the unit melts or burns up. With a resistor-type speed control, the more often partial throttle is used, the hotter the controller will become. To reduce problems with these types of controllers, use partial throttle as little as possible.

To further reduce the chance of having heat-related problems with your mechanical control, make certain that there's no nearby plastic that could melt or burn if the controller gets too hot. This includes hull parts and any wiring or radio cases. Move the controller to a better position if necessary, or shield the plastic with pieces of sheet aluminum. Mount the aluminum about ¼ inch away from the plastic; that should protect it enough, except from the most severe heat.

Entry-level boats usually come with mechanical controls. But when, to improve performance, you modify your boat with a larger prop, a hotter motor and better batteries, the usual mechanical control is living on borrowed time. If you run your fast electric boat often, or you race it on a tight course, chances are very good that you'll eventually destroy

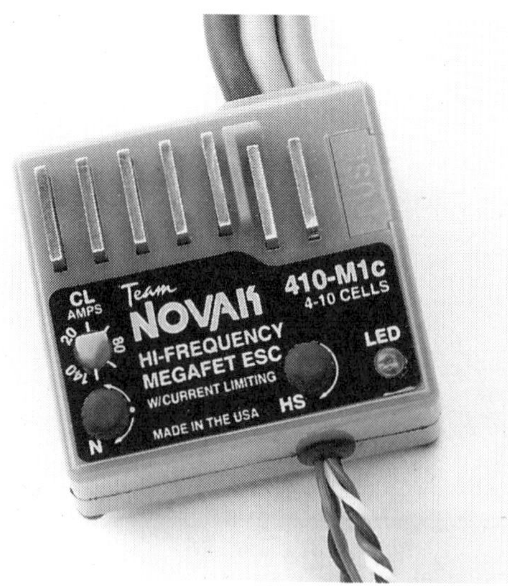

As long as it's kept dry, an R/C car ESC such as this one by Novak can work really well in a boat. A watertight radio compartment or ESC box is highly recommended.

your controller. One of the first investments a racer should make is in a high-quality ESC.

There is one exception, however. The mechanical speed control made by Hughey Boats is significantly different from other mechanical types. It was designed to handle high current loads when racing; other mechanical controls were designed to be an inexpensive method of varying boat speed. The Hughey controller still produces heat at partial throttle, but it was designed to handle it. Using this mechanical control instead of an ESC is not a compromise; many speed records are held by boats running with this Hughey product. If you can't afford a top-quality ESC or think they're too complicated, the Hughey unit is well worth considering.

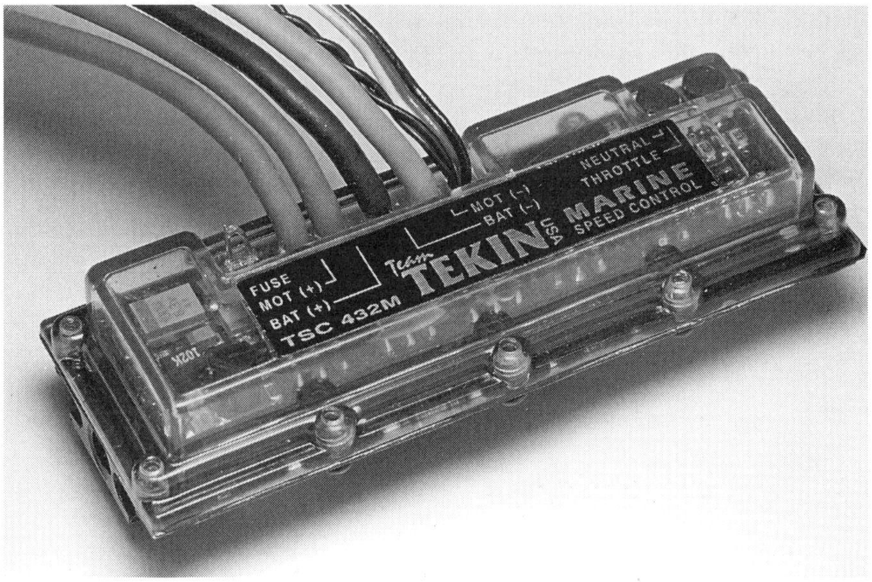

Tekin's* 432M marine controller will handle up to 32 cells and two motors, and it can be water-cooled—all while being waterproof (follow the directions carefully). Current models perform very well and are recommended instead of an average R/C car ESC. If you don't intend to run more than 14 cells, you don't need an external battery pack.

ELECTRONIC SPEED CONTROLS

Most of the ESCs used today are of the high-rate type that gives smoother control and more efficient operation. Simply, an ESC slows a motor by rapidly switching the current on and off thousands of times a second. Before investing in an ESC, evaluate your motor and battery system and estimate your current requirements. Most fast electric boats require a minimum rating of 200 amps continuous current for high amp draw and low ESC resistance. Check with an ESC's manufacturer for suggestions on its use.

Try to estimate the number of cells you'll use with the ESC, too. True, today, you may only run a 7-cell system, but next season, you may want to run 12 cells, and many ESCs won't handle that many. Save money by buying a more flexible ESC today and using it with a different cell count tomorrow.

Choose the best ESC you can afford, but base your decision on features you actually need. Current limiting might be useful in some circumstances, but you certainly don't need brakes! Go with the ESC that has the lowest "on" resistance and will handle the cell count you need and the current load you expect (generally, no less than 200 amps).

I've successfully used the popular R/C car ESCs; if properly protected from water, they work really well. Special marine ESCs differ from car controllers by not having reverse or brakes, by having ratings for more than 12 cells and, often, by being waterproof. Your choice should be made according to features, cost and availability.

It may seem elementary, but follow the instructions that come with your ESC very carefully. More ESC problems arise from not following these directions than from any other single cause. One of the most common mistakes is not installing the required motor capacitors. Most modern ESCs need three on each motor: one on each lead going to the motor can, and one connecting the two leads. Failure to install these as required can result in radio problems, and you might even damage the ESC.

Some modelers find that adjusting the ESC's control potentiometers (pots) is tricky, but if the correct radio settings have been made, the ESC can be adjusted. Most instructions call for the idle pot to be adjusted first. After the pot has been set so that the motor does not turn over with the throttle in the off position, advance the throttle to about 80 to 90 percent of full movement, and set the high-speed pot to maximum. If the high-speed adjustment is made with the transmitter throttle control fully against the stop, it's possible that the last bit of full throttle in the ESC will be lost. By setting the maximum adjustment with the throttle short of full "on," you'll be assured that the ESC will deliver maximum power to the motor. Some of the newest ESCs are self-setting; they're correctly adjusted with just the push of a button. I hope that this trend continues.

The ESC instructions should include a chart showing the transmitter adjustments that you have to make before you adjust the ESC itself. Be certain that you have the transmitter adjustments, including throttle reversing and throttle trim, set as directed in the instructions for your particular ESC. ESCs don't all need the same transmitter adjustments. If the transmitter settings aren't correct for your ESC, you may not be able to adjust the ESC correctly, and it won't work properly. Before you install an ESC, take the time to read and understand the instructions, and you'll be much happier with the results.

Astro Flight's Hydro-Racing ESC is very popular with top racers, and it holds many speed records. Its only major drawback is the absence of a BEC feature; other Astro ESCs have BEC.

• **Reverse battery polarity.** If you hook up an ESC with the battery polarity reversed, you'll have problems. Some ESCs will be badly damaged and need factory repairs; others have fuse protection and will work again after being reset. Avoid this problem by using polarized battery connectors, which can't be hooked up backwards, and be careful when working on the boat's wiring.

• **Overheating protection.** Many ESCs have overheating protection that will shut the unit down if it gets too hot. If your boat mysteriously stops after running for a while, check the connectors and wiring. If nothing seems amiss, the ESC may have shut itself off. Wait a few minutes and try the throttle again. If the motor works, the ESC probably overheated. If you don't want to eventually damage the model's electrical system, you'd better solve the overheating problem.

• **Waterproofing**. Water is the enemy of all ESCs, so take the time to either waterproof the unit or install it in a waterproof compartment. The easiest way to do this is to seal the entire hull with a waterproof deck. A wet ESC probably won't work, and if left wet, it will be damaged. If yours gets dunked, take it home and, if possible, open the case. Use a hair-dryer to heat the components so that they dry completely. Don't get carried away and fry the ESC with the dryer; warm the components; don't get them hot.

• **ESC failure.** If your ESC fails to work, you'll probably have to return it to its manufacturer for repair. Before you send it in, it might be worth calling their help line to see whether they can suggest something you could try. In truth, most ESCs are very reliable and will function well for years if they aren't abused. Keep them dry, use the motor capacitors, don't hook them up backwards, and don't run more current through them than they were designed for, and they'll last for a long time.

Two monos round a buoy onto the back straight. You need some kind of speed control to slow the boat down for the turn; otherwise, you risk a spinout.

CHAPTER 7

Hardware

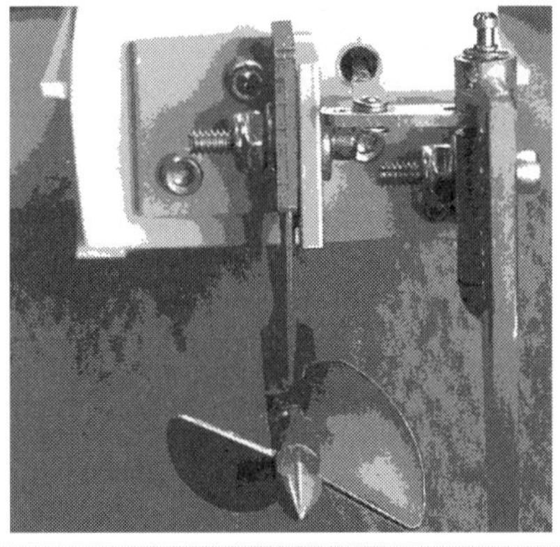

Two hydros round a turn buoy, throwing large rooster-tails as they accelerate into the straight. Drive-line setup and design greatly affect a model's turning capability.

DRIVE LINES

A solid drive shaft is the traditional marine drive system, and it has much to recommend it. It's simple, relatively cheap and rarely breaks. Its disadvantages are that adjustment possibilities are limited, and it's very sensitive to poor shaft/motor alignment.

The typical fast electric solid drive system is an ⅛-inch steel shaft connected to the motor with either a solid coupling or, better yet, a universal joint. The U-joint is much preferred for two reasons: it compensates for some degree of misalignment in the installation, and it allows some momentary misalignment while running. Any misalignment causes drag and a loss of power to the propeller, so it must be minimized or avoided if possible. Even a perfectly aligned installation with a solid connector will encounter drive-line drag when the hull flexes and the motor and shaft move independently of one another. This phenomenon is common, especially with plastic hulls. A small U-joint will compensate for this momentary misalignment, and there will be only minimum power loss.

The shaft must be straight of course, and this can pose a problem for the scratch-builder. Most of the music wire sold for shafts is not straight, and using it will decrease performance because it increases drag and vibration. Try to obtain drill rod or some other material that's guaranteed to be straight. If in doubt, chuck the shaft material in a drill and spin it slowly; bends will be easy to see.

The shaft is supported in a stuffing box that's usually made of ⅛-inch-diameter brass tube. This works well in most cases, but to prevent shaft whip, it's best to avoid exposing more than 3 inches of the shaft. The shaft is lubricated with light machine oil, and to prevent it from rusting, it must be removed after

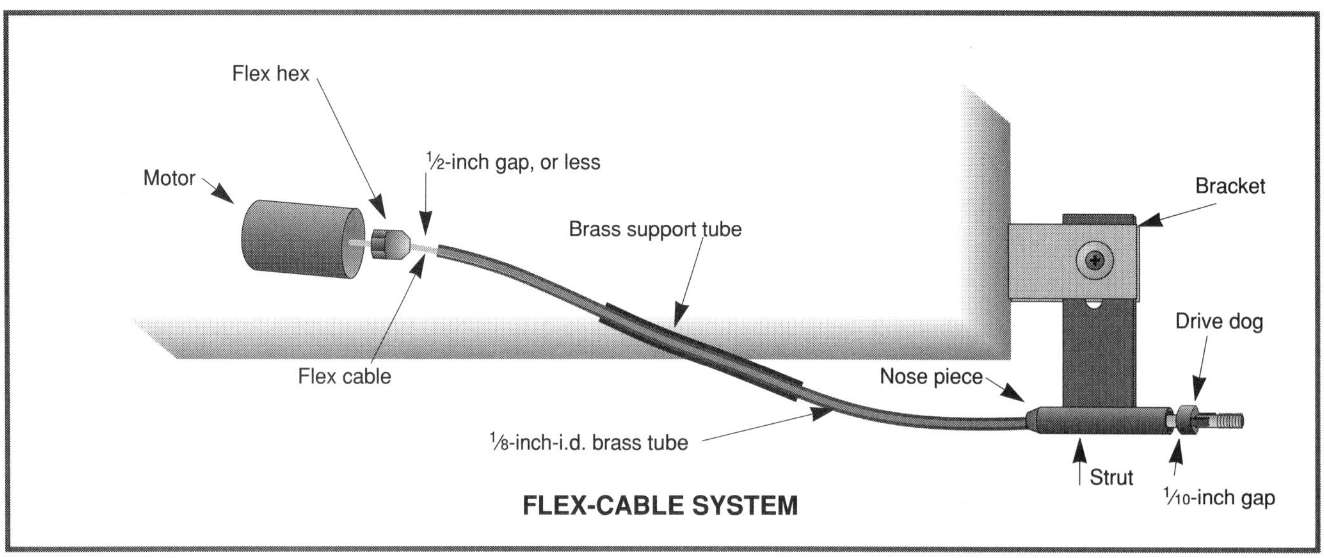

A cross-section of a flex-cable-drive setup. The cable tube should be inserted into the strut's nose piece. The gap between the strut and the drive dog is essential to allow for cable shortening caused by wind-up and thrust transfer.

each outing and dried off before it's re-oiled.

The diameter of the drive shaft should suit the power output of the motor and batteries. Six- to 8-cell boats running single 05-type motors usually run 1/8-inch shafts. Larger, 12-cell boats running larger motors or geared drives will usually run 3/16-inch shafts. The larger shafts fit the larger propellers usually used on the 12-cell boats; these have 3/16-inch holes for 3/16-inch drive shafts.

You may occasionally want to use a prop that's too large to fit on an 1/8-inch drive shaft. Some racers like to use the Octura X437/3 prop on light hydros running the older five or six-turn Astro Flight motors. This prop is designed for a 3/16-inch shaft, but it can be installed on an 1/8-inch one by using short lengths of brass tube to bush the shaft larger. This works well, but you must also substitute a 3/16-inch drive dog for the 1/8-inch one. File a hole in the side of the brass tubes for the drive-dog setscrew, and use a washer and nut to hold the larger prop on the bushed, 1/8-inch shaft.

Even a solid shaft running in a fixed brass stuffing box has some adjustment possibilities. On a hydro, if more lift is needed, moving the prop aft on the shaft will position it deeper in the water and help to raise the stern. Conversely, the stern can be lowered by moving the prop forward on the shaft. Of course, this only works if you planned ahead and made the shaft long enough to allow this type of prop adjustment. You can sometimes loosen the motor-coupling setscrews and pull the drive shaft aft a little to allow more adjustment, but the best solution is a longer shaft.

THE FLEX SHAFT

In very fast electrics, the most common inboard drive system is the flex shaft. This consists of a length of flexible cable (such as automobile speedometer cable) that's soldered to a short piece of solid shaft. With this system, the solid shaft rides in bearings mounted in an external strut while the cable runs through a tube to the motor. A coupling connects the cable to the motor.

Two types of coupling are common: a hollow metal bar with two setscrews, and a collet-type coupler such as Octura's Flex-Hex. The bar type is generally less than satisfactory, and its use often results in the loss of the entire drive shaft and propeller. The exception is the rather robust coupler offered by Hughey Boats. This large-diameter coupler has long setscrews, and it holds up well.

Still, for most uses, the collet-type couplers are probably best. They will grip even an oily flex shaft, and if tightened sufficiently, they will not loosen. It's possible to overtighten them, but common sense should guide most modelers. One tip: to prevent the flex shaft cable from unraveling, many modelers tin the motor end with solder. This works well, but if the solder is built up too thickly, it will prevent the flex coupling from being tightened adequately on the cable. To avoid losing the cable and prop, be sure to sand the solder down flush with the cable's surface.

Most direct-drive boats use a 0.098-inch-diameter cable that's soldered into an 1/8-inch stub shaft. This allows the use of the common propellers with 1/8-inch holes. With gear drives or large motors though, a heavier, 0.130-inch-diameter cable is often used. This cable is soldered into a 3/16-inch stub shaft that fits the larger props with 3/16-inch holes. The larger cable is stronger, so it won't break under the heavier loads imposed by 12 cells and dual motors.

A typical surface setup with direct cable drive. The Octura strut and rudder are held on brackets made of angled aluminum, and plated bolts hold everything together. Note the prop's downward thrust line; it helps to lift the transom.

SOLDERING SOLUTIONS

I mention soldering the flex cables into the appropriate stub shafts, but this causes many boaters problems. I've rarely had trouble with soldering, but I was amazed at how often club members lose stub shafts because their solder joints break. I asked a few questions, and the cause of the problem became clear: many boaters were using electrical solder to attach the flex cable and the stub shaft.

Electrical solder is *not* strong enough to withstand the torque of a powerful fast electric motor spinning a big prop. If you solder your flex cable, use *only* silver solder and a propane torch; but even this strong solder will fail if you can't make a good solder joint. A related problem is that the flex cable may be overheated and weakened when it's being soldered.

If you don't have much confidence in your ability to solder well, find someone who *does* solder well. Pre-soldered cables are available from Hughey Boats and others, and they may be a good idea for some people. Remember that a poorly soldered flex cable will not only let you down at some point, but it will also lead to your losing an expensive prop!

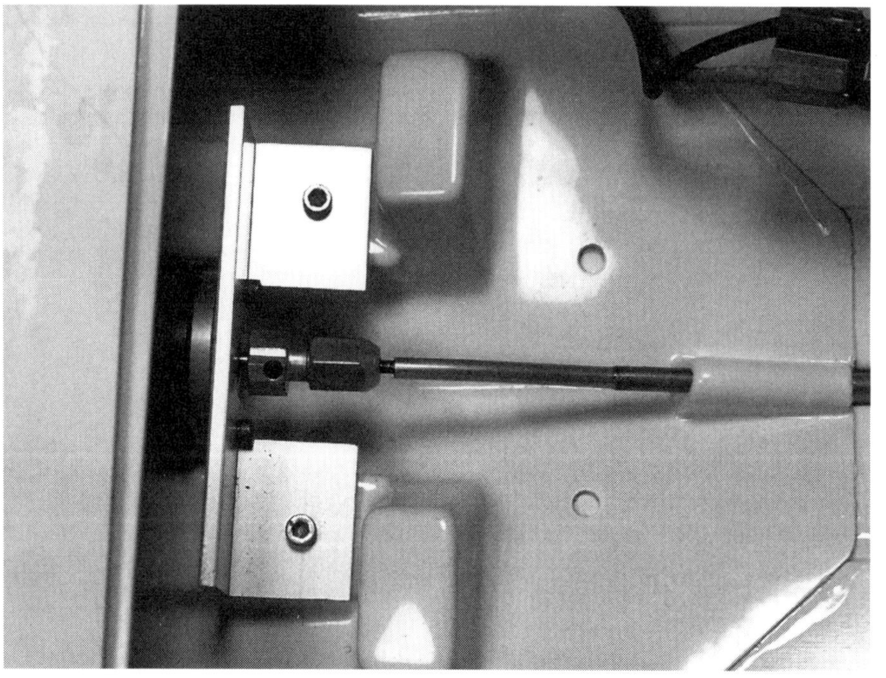

This simple but robust installation works well. The motor mount is made of angled aluminum, and the flex cable runs through an 1/8-inch-i.d. brass tube rather than the usual Teflon™ tubing. The brass supports the cable better, and if good waterproof grease is used, there's virtually no wear inside the tube. Note the small gap between the end of the stuffing tube and the flex-hex coupler; this prevents "cable whip," which can occur if the gap is larger than 1/2 inch.

STRUTS

Struts hold the propeller shaft under the boat. Some of the solid drive shafts can get by without a separate strut, but almost all flex-cable drives need them. One of the big advantages of a flex drive is that it can be adjusted to help tune the boat's running attitude and handling; the strut is the adjustable part of the system. Most struts are mounted on the transom with an adjustable mount. By loosening a bolt on the mount, you can move the strut up or down or change its angle.

Many modelers use struts made out of brass sheet and tube, while others use cast struts such as those marketed by Octura. Octura struts are cast in both aluminum and beryllium copper; the aluminum is recommended only for smaller, direct-drive boats. Other manufacturers offer struts made of strengthened plastic or sheet metal; the choice is yours, but strength and intended use should be your prime considerations. For optimum performance, minimize drag by using a strut that has a small cross-section, but with some boats, be sure that it's strong enough to withstand speeds of 30mph or more.

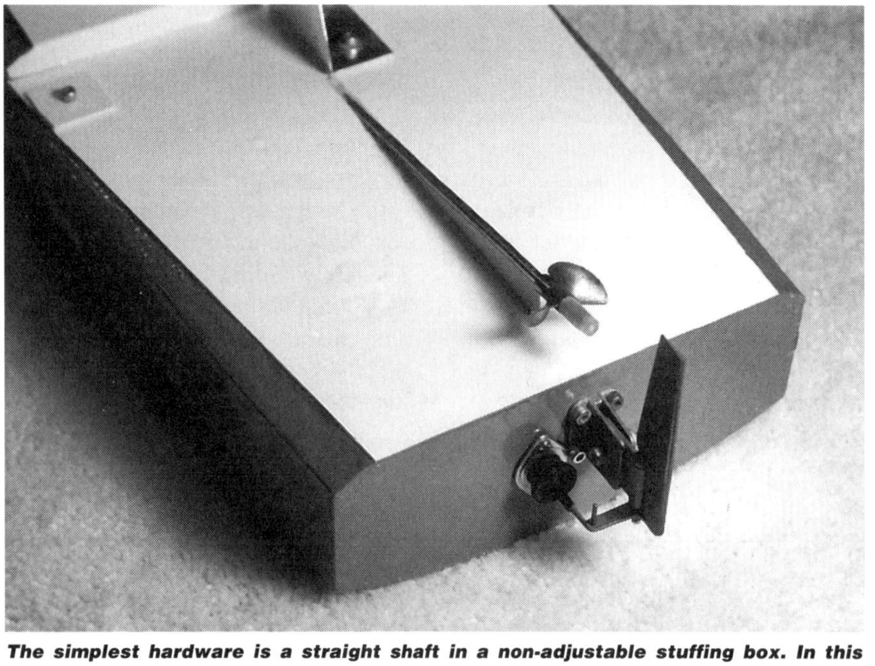

The simplest hardware is a straight shaft in a non-adjustable stuffing box. In this hydro, a Bru Line* waterproof seal on the rudder pushrod keeps water out of the interior.

RUDDERS

In a typical scale model boat, the rudder will be mounted under the hull behind the prop. For fast electric use, the rudder should usually be mounted on the transom, but it doesn't have to be set directly behind the prop. A fast electric boat's rudder will be made of materials such as those used for struts, and to minimize drag and maximize speed, low weight and small cross-sections are important.

A good, high-performance rudder will be wedge-shaped rather than the more conventional streamlined shape. The wedge shape causes less drag for this reason: as the rudder cuts the water, the vacuum created behind it is filled with air instead of water. Air will fill the vacuum faster than water, and the net result will be less drag and better turning.

Mounting the rudder on the transom allows it to be positioned farther aft for better turning. If an adjustable mount is used (and it's highly recommended), the boat's turning characteristics can be changed to suit the model and the water conditions. Tilting the bottom of the rudder toward the boat will raise the bow in turns and can help to prevent it from hooking and spinning out.

It isn't always advisable to mount the rudder directly behind the propeller. For maximum maneuverability, slower boats and boats that must negotiate a very tight course need the extra steering provided by the propwash running over the rudder's surface; but on a fast electric boat, the boat's forward speed is enough to load the rudder properly. On a fast boat with a surface-drive prop, mounting the rudder directly behind the prop means that it runs in "dirty" water full of bubbles. This can make the boat wander in the straights, because the rudder is less than fully efficient. Especially on hydros, the rudder is often mounted to the right or left of the prop, just as it is on full-size inboard hydros. As long as the model is running at high speed, rudder control is excellent.

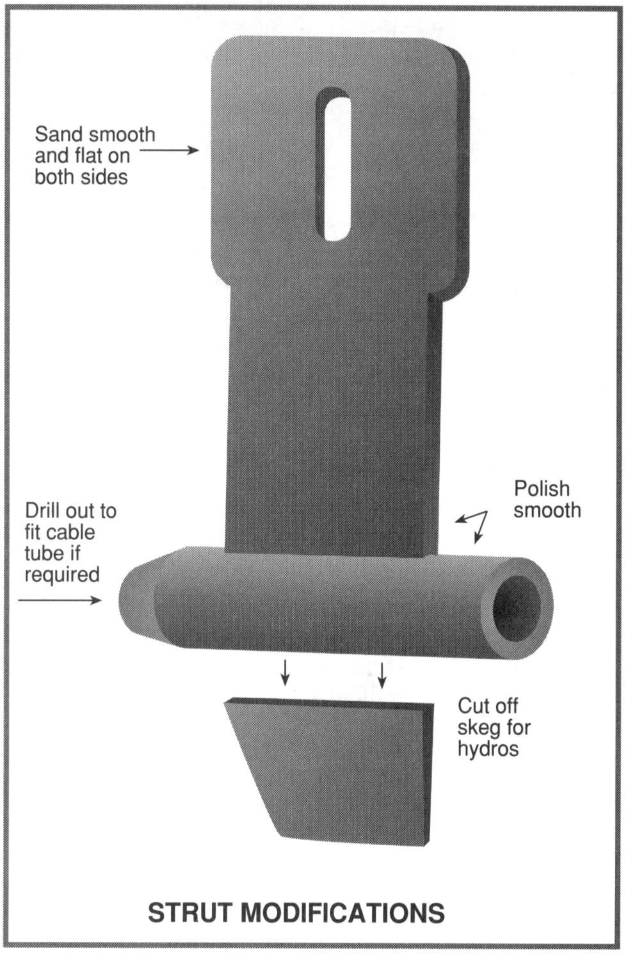

A cast strut may need some work before it finds a home in your fast electric boat. A three-point hydro should not have a skeg below the bearing tube; cut it off.

FINS AND TRIM TABS

Because a fast electric boat running at top speed is not always the most stable thing on the pond, extra control devices must sometimes be added.

- **Turn fins.** These are probably the most common fins; they're used to provide a surface for the model to pivot against when it turns. A shallow-vee mono may slide too much in a corner; adding a turn fin to the keel just ahead of the balance point helps to anchor the front of the boat so that the transom can come around in a turn. On a hydro, a turn fin mounted on the inside spon-

The layout of this Graupner outdrive is typical of most outdrives; the rudder is directly behind the propeller. The rather large assembly causes more drag than an all-metal outdrive, but on a short course with tight turns, the Graupner setup has been proven to reduce lap times.

son serves the same purpose: it prevents the model from skidding straight ahead when the rudder is turned.

• **Spin fins.** As its name implies, it prevents a spinout by increasing a boat's resistance to sliding. Usually mounted on the inside of the transom, in a turn, the spin fin keeps the rear of the boat behind the bow and reduces the boat's tendency to hook. The spin fin should be mounted so that it's at a right angle to the water's surface when the boat is in the turn; if it isn't, it may actually lift the transom and cause a spinout. (On a deep-vee, for instance, mount the fin perpendicular to the hull bottom, because the boat will heel inward in a turn.)

• **Trim tabs.** These control a boat's ride attitude. Especially useful on boats that don't have an adjustable drive, the trim tab is mounted on the transom to allow the boat's ride attitude to be changed. If a boat's bow rides too high, adjust the trim tabs (down) to lower it. If mounted near the chines, trim tabs can modify the way a model corners by keeping it flatter in a turn.

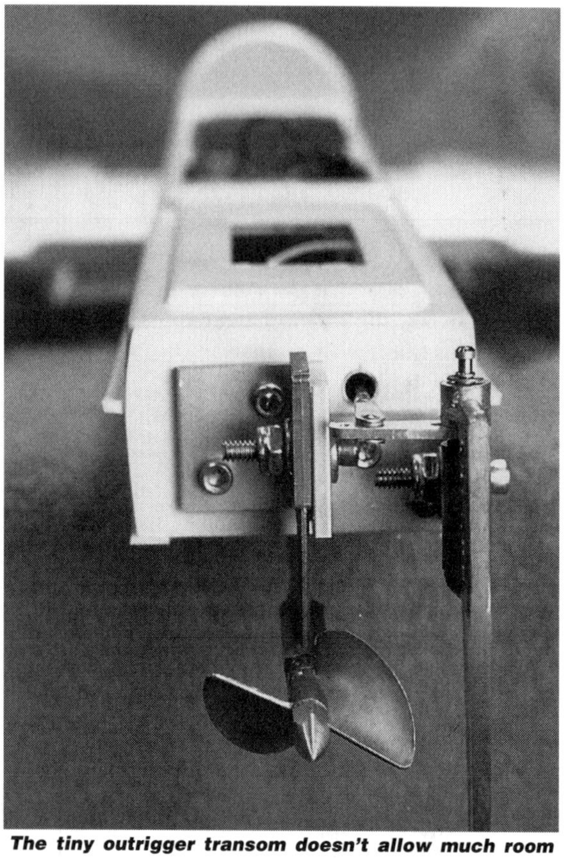

The tiny outrigger transom doesn't allow much room for hardware. The rudder is commonly offset to the left or right of the prop to allow it to run in undisturbed water.

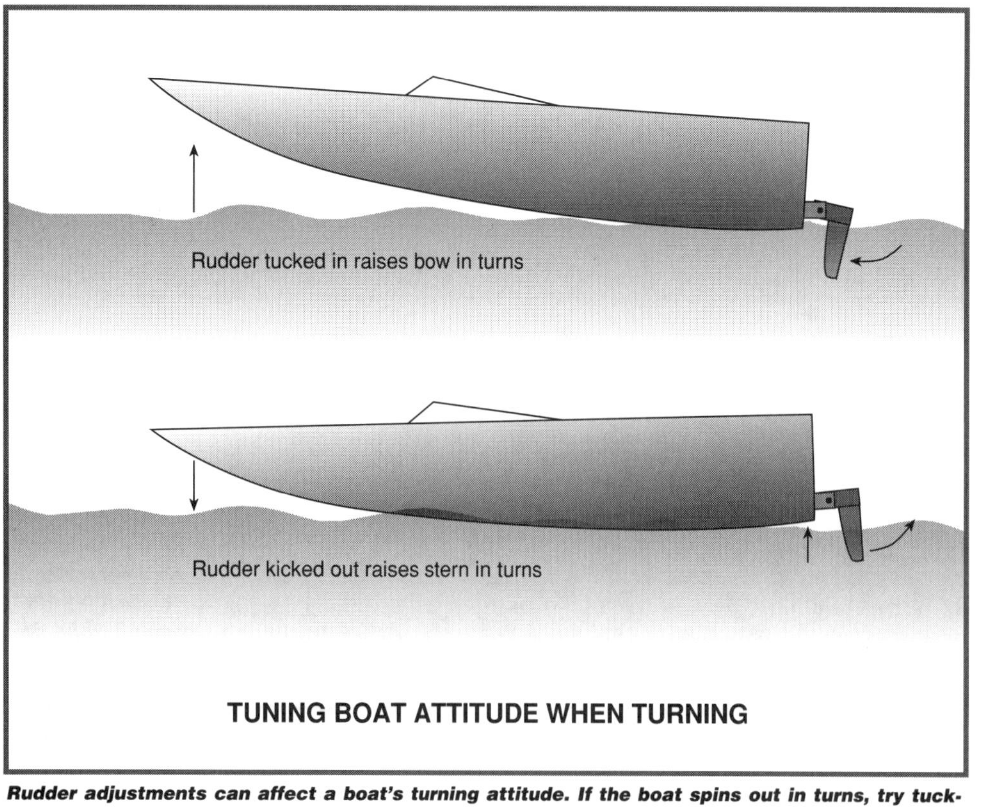

Rudder adjustments can affect a boat's turning attitude. If the boat spins out in turns, try tucking in the rudder a few degrees. This will raise the bow and anchor the transom. Rudder kick-out is seldom used on monos because it promotes hooking, but some hydros will benefit from the transom lifting.

CHAPTER 8

Propellers

THEORY AND TERMINOLOGY

The propeller is one of the most important pieces of equipment on a fast electric boat, and it's one of the most misunderstood. Any boater can tell you that the prop pushes the model forward, but few can accurately describe how it does this. Read on, and you'll learn enough propeller theory to impress your friends and make your boat go faster!

Most people compare a propeller with a screw; if a wood screw has a thread "pitch" of 10 threads per inch, for each turn, the screw will advance 1/10 inch into the wood. Similarly, many people reason that a boat propeller behaves in the same way by advancing a fixed distance through the water for each revolution. This is partially correct, and it serves as a useful model of propeller function.

One major difference between a wood screw and a propeller is the slippage factor. As a propeller spins under water, it will "slip," e.g., a prop with a pitch of 2 inches will actually move forward less than 2 inches per revolution. The slippage is usually expressed as a percentage: a 2-inch-pitch prop with a 50-percent slippage will move forward 1 inch per revolution.

This slippage varies greatly and depends on the model, its powerplant and the speed at which it's running. Obviously, if you hold your model still in the water and apply full throttle, the slippage will be very high and the boat won't move forward. A lightweight hydro flying at full speed might have slippage that's as low as 10 percent, while a sport mono on plane might slip 25 percent. Though we want to minimize slippage, some is required to provide thrust for the boat. For the average fast electric boat, slippage probably ranges from 20 to 30 percent.

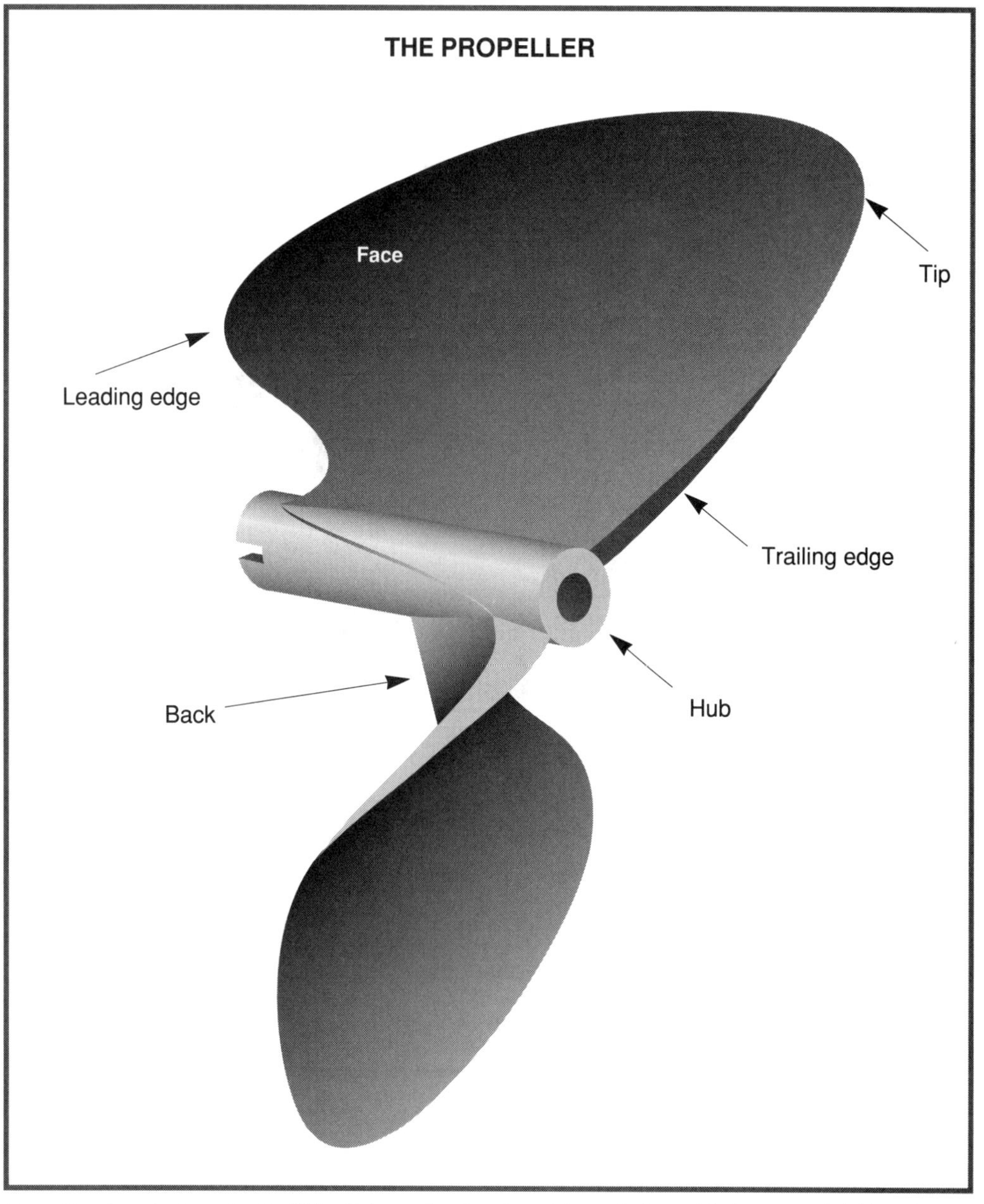

Propeller terminology isn't difficult to understand, and knowledge of it is essential if you're to understand how propellers work.

Most manufacturers list their props by a stock number only, and a reference chart is usually needed to determine the two numbers in which we're most interested: the *diameter* and the *pitch*. We need to know these numbers so that we can choose the correct propeller or propellers for our fast electric model.

- **Prop diameter** is the maximum measurement across a prop's blade tips, and it's important. The larger a prop's diameter, the more water it will push. This can be visualized as a "thrust cone," i.e., a cone of water that's moved backward by the spinning prop (see page 18). The larger the prop, the larger the thrust cone and the greater the volume of water moved per revolution. A larger prop requires more motor power, and if it's too large, the model will need either a gear drive (to avoid overloading the motor) or a larger motor.

The thrust cone can be considered as a kind of traction value. The larger and heavier the boat, the larger the thrust cone must be to push it adequately. Smaller models (20- to 25-inch beam) can usually get by with 1.05- to 1.4-inch-diameter props, while larger boats may need props of 1.4- to 1.8-inch diameter. A lot depends on the model's type and weight, but my figures can be used as a guide.

• **Prop pitch** is the distance a prop would travel per revolution in a solid medium.

CHOOSING THE CORRECT PROP

Most boat kits come with a prop and many also come with motors. Most manufacturers, however, supply low-performance motors and props that provide long run times at low amp draw. This is the best approach, because such a setup is less likely to frustrate beginners, and advanced modelers will be able to hop up their models with aftermarket props and motors. I've often turned a relatively low-performance model into a "tiger" by changing certain components.

Most kit boats respond well to a prop change, as long as you don't get carried away. First, try a prop with a slightly higher pitch, then try one with a slightly larger diameter. Run the boat at full speed for 1 minute with the new prop, then check the motor's temperature. If the motor is too hot to touch, the new prop is too big; try one with a smaller diameter or a lower pitch. You may have to modify the model to take aftermarket props such as those in Octura's extensive line but, in many cases, e.g., Graupner and Kyosho boats, the manufacturer offers several props that will fit. Graupner and Octura both make adapters that allow the use of Octura props on the metric prop shafts found on many kit boats.

If you've already built your boat or you've modified your kit boat, choosing a propeller is more difficult. Replacing the kit's motor with a hot modified one will almost always require a prop change. If the prop has been optimized for the stock motor, in most cases, you'll have to fit a smaller prop with the hotter motor, because the hotter motor draws more amps at low rpm than the stock one. Loading the new motor down with a prop that's too large will make it run too slowly and overheat. Choose a smaller prop that allows the new motor to wind up to higher rpm. This will produce higher speeds because the smaller prop will be turned at much higher rpm.

This direct-drive mono has a non-adjustable surface-drive prop—an Octura X435. The X Series props work especially well on monos because they produce very little lift, which is a cause of hooking. The prop isn't held by a prop nut; tightly fitting fuel tubing works better.

This problem with prop selection surfaced in my local club when several members raced Kyosho Vipers in a special Viper class. Hoping for more speed, several racers had replaced the kit's motor (a potent stock motor in its own right) with a hot R/C car motor. One club member installed a 16-turn modified, but his boat was only slightly faster than it had been with the Kyosho motor, and his run time was less than 1 minute. He often failed to finish a race because his motor or ESC overheated. When he replaced the large Viper prop with an Octura X430, his speed increased, and his run time lengthened to more than 2 minutes.

A variety of surface-piercing propellers can be carried in a plastic fishing-lure box. If you spend hours balancing and polishing your props, you should protect them in a suitable container.

The only way to know which new prop will work best is to time the boat over a measured course using several props. Run your boat around two buoys (bleach bottles) while someone else times the run with a stopwatch. Using the old prop, get times for several laps, then bring the model in, change props and install a new battery while the motor cools down; then time the boat with the new prop. Be sure to carefully record all your test data!

When you've found the prop that gives the lowest lap times, test it for run time: let the motor cool down completely; then install a fully charged battery pack; time the model around your test course until the battery dumps (the model slows down rapidly). If you use a 1400mAh battery, you should have at least 2 minutes of run time; a 1700mAh pack should run for almost 3 minutes. These times indicate an amp draw of 35 to 40 amps, which is about as high as you'll want to go unless you're racing. If the battery dumps sooner, choose a slightly smaller prop.

Some propellers just won't work on certain models. Here, a hydro with a surface-drive setup tries to plane off with a 1445 prop that doesn't produce enough thrust to get the model moving. An X445 prop, which is designed for surface drives, gave immediate acceleration and good top speed.

Though the propeller's diameter and pitch are the most important factors for prop selection, other characteristics should also be considered. The most popular aftermarket props are those in the Octura X Series. They're versatile enough to work well in almost any application, and I recommend that beginners try them before using others. The X Series low-lift props work submerged and in surface applications, and they usually give good turning performance. Low-lift props direct most of their thrust to the rear and do not tend to lift a boat's transom at speed. Other manufacturers offer props of similar design.

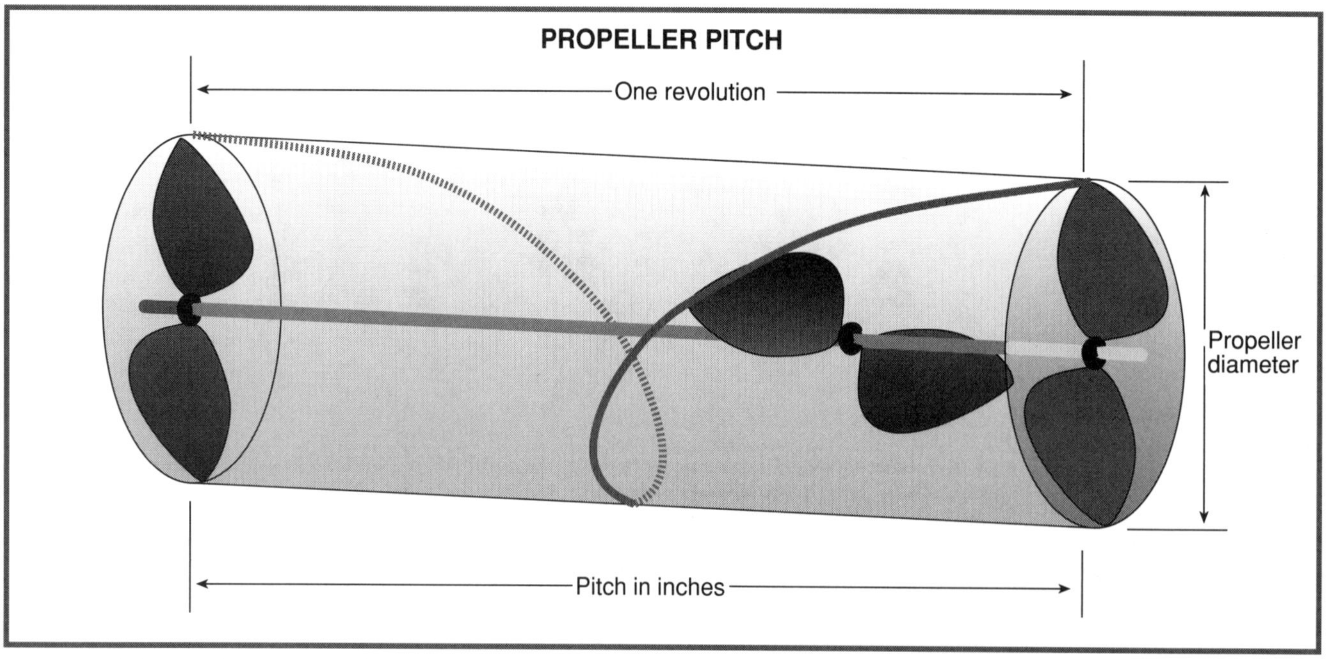

A prop's pitch determines how fast it must spin to reach a certain forward speed. The higher its pitch, the faster the model. (A prop with a higher pitch requires more power to turn it.)

Some boats may do better with a change of prop. One obvious example is the three-point hydroplane. At full speed, a hydro contacts the water at only three points: the two front sponsons and the prop. Some hydros drag their sterns when they run a low-lift prop, and the extra drag really slows them down. A change to a lifting prop can help to get the transom into the air and can really increase a hydro's speed. You could, for example, replace an Octura X640 with a 1640; their blades are shaped differently, but they have identical diameters and pitch. The 1640 is a lifting prop, and it might be just the ticket for a hydro.

Many boaters encounter the problem of cavitation. It occurs when the prop can't get a good "grip" on the water and spins inside a pocket of air. Little or no thrust is produced when a prop is cavitating, so the boat barely moves. Cavitation is often caused by running the prop too shallow, but the problem usually results from the prop's design. The X Series props usually "throw off" the bubble so that the boat accelerates well; other designs may not work at all from a dead stop. If, at full throttle, your model just sits there and sprays water, try a more gentle start. If that doesn't work, try tossing the boat forward while you hold down the throttle. If the boat hits the water prop first, it should accelerate away; if it doesn't, you may have to try a different prop.

Its blade design determines a prop's use and performance. Left: this X Series prop is a good all-around design that works well in almost any application. Right: a 1700 Series prop that will perform better than the X Series in some instances. One of my hydros picked up 4mph when I swapped its X Series prop for the 1732 shown here.

Metal props must be balanced before they're used in high-performance fast electric boats. Certain model aircraft prop balancers will work with boat props, but this Octura balancer is better than any other I've tried.

PROP MATERIALS

Model boat props are usually made of plastic, aluminum, or beryllium copper. For sport and for some racing, plastic props work well, if you can find the right one for your model. Contrary to the advice of some "experts," the prop blades shouldn't flex when in use; the stiffer the blades, the better.

• **Plastic.** To prevent the blade from flexing, the best plastic props use graphite or a similar stiffening fiber as reinforcement. The advantages of plastic props are their low cost, and they're usually well-balanced.

• **Aluminum** is good for propellers. It's light, which almost eliminates the need for balancing, and it's easy to polish and re-contour. Although aluminum is stiff and won't flex much, it can crack if bent, and it costs the same as props made of harder materials. Aluminum is too soft to produce a sharp leading edge, and a sharp blade entry is necessary for maximum performance at very high rpm.

• **Beryllium copper** is often used to make very strong, stiff, high-performance propellers that won't break easily if bent. These props can be sharpened to a razor edge and polished nicely, but they will almost certainly require balancing. Most important, propellers made from this and similar materials are the choice of virtually all serious fast electric racers, and most records are held by boats that run with beryllium propellers.

PROPELLER BALANCING

The props used in fast electric boats can turn at from 10,000 to more than 25,000rpm when the model is at top speed. When light plastic or aluminum props are used, vibration is seldom a problem because the props are so light that they can't be severely out of balance. An unbalanced beryllium or bronze prop is a different matter; a prop with a heavy blade turning at more than 20,000rpm can produce enough vibration to cause radio problems or cracks in the hardware or in the hull itself. For high-performance use, *no* propeller should be run without first being balanced.

A properly balanced "wheel" requires a good balancer. A simple one can be made of scrap wood and two razor blades, but if you work on many props, it's probably best to buy a commercial balancer. I've used several types, but the best are those specifically designed for boat props. Some of the airplane prop balancers are not sensitive enough for a small boat prop. As I write this, the balancer I like best is the one offered by Octura. It's simple, rugged and can easily be adjusted.

To balance and finish a prop, you'll need:
—a 10-inch mill file;
—180-grit and 400-grit wet-or-dry emery paper;
—a hand grinder with a carbide-cutting wheel and a polishing bob. (This is handy, but it isn't necessary to do a good job.)

Put the prop on the balancer, and mark the heavy side with a marking pen. Remove metal only from the back of the blade, or from the parts that face the bow. Start with the blade's leading edge, then move slowly toward the hub, checking the balance often. It will take some practice, but you'll soon have a good feel for how much you have to remove to achieve a balanced prop. When the blades are almost perfect, use wet emery paper to remove scratches and obtain a smooth finish.

Wet 400-grit paper will produce a finish that's good enough for any fast electric use, but the surface can also be buffed to a bright finish. There's little evidence to indicate that a high shine means a faster prop, but many boaters like the look of a bright propeller. If you power-buff a prop, be very careful: don't cut yourself on its sharp edges. And be careful not to round off any edges or to polish hollows in the blade's face or back.

***Important note:** when you work on a beryllium prop, always use a good respirator and goggles to keep the fine metal particles out of your lungs and eyes. Beryllium is poisonous! Please invest in good protective equipment!*

This 12-cell hydro throws a large rooster-tail off its surface-drive prop as it accelerates out of the right turn. On an oval course, models like this hit speeds of more than 40mph.

MODIFICATIONS

If the right prop just isn't available, you can modify an existing one to suit your needs.

• **Reducing blade area.** This is the most common modification. If your motor won't pull a stock 1732 prop, you can reduce the blade area slightly to allow it to turn up higher. The popular "Texas cut" accomplishes this by trimming away part of the leading edge of both blades. Another modification involves cutting the blades to the contour of a prop with a smaller diameter. This results in a prop with higher pitch than a smaller stock prop, and the new contour may allow the model to go faster.

• **Blade cupping and re-pitching.** These are beyond the scope of this book. Custom prop builders are often listed in the advertising section of model boat magazines, which are also good sources of additional props and information.

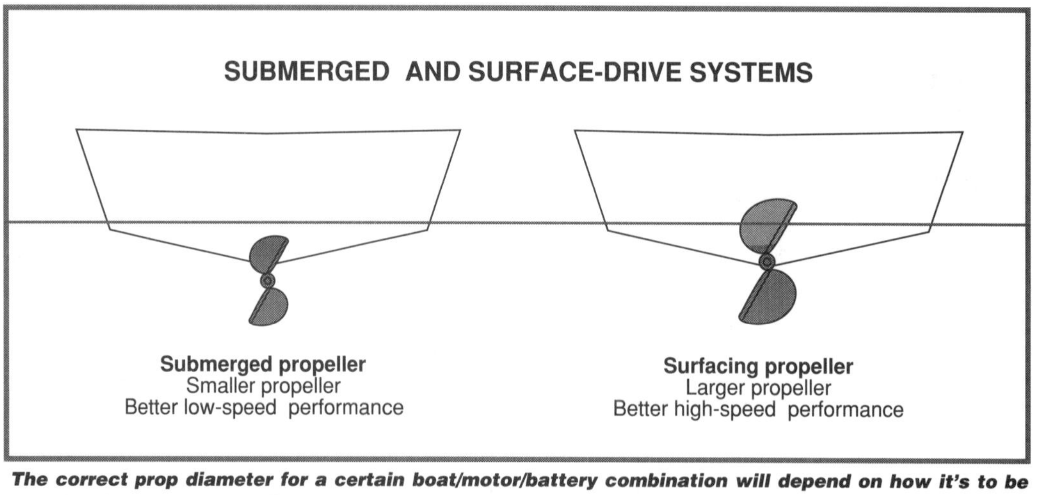

The correct prop diameter for a certain boat/motor/battery combination will depend on how it's to be run. A submerged prop offers better low-speed performance, but the increased drag on the blades requires a smaller diameter than if it were a surface drive. If the blades are out of the water part of the time, the load on the motor is reduced, and a larger prop can usually be installed.

CHAPTER 9

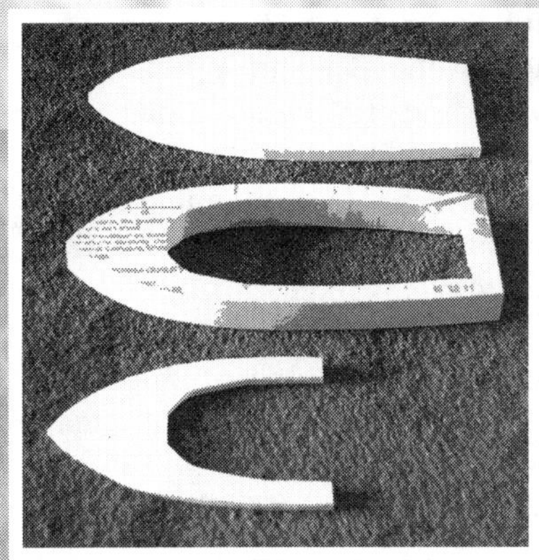

Construction

PLASTIC HULLS

When my interest in the R/C hobby began years ago, most electric boats had wooden hulls built from kits. There were a few smaller, fiberglass hulls, but they were designed for glow power and were rather heavy. In contrast, most of today's fast electric boats are based on commercial plastic hulls. These have many advantages: they're light (usually), are available in more complex shapes, have more scale appeal, take less time to build and have a superior finished appearance. They may also cost more than a similar boat built of wood, but if lightness and the resulting superior performance are your goals, a plastic hull is a good way to go.

From a building standpoint, the ABS plastic used for most vacuum-formed hulls is pretty user-friendly. CA glues will bond very well to it, but some epoxies will not. Most paints will stick well as long as the plastic's shiny surface is scuffed first.

Most plastic kits contain vacuum-formed parts: generally, a two-piece hull with a removable deck and, often, other decorative parts such as wings and crew figures. The thickness, therefore the strength, of the plastic varies from manufacturer to manufacturer; the thinner plastic is lighter yet weaker. Obtaining a compromise between strength and weight is always an issue, and the ultimate use of the model is a major determining factor.

CA glues are a boon to boat modelers because they speed construction and, in many cases, give superior joints. Buy a bottle of thin glue and gap-filling glue and a spray bottle of accelerator. Practice with them first; and always be careful and wear safety glasses!

USES OF ABS

Graupner's little "Key West" is arguably the fastest direct-drive, readily available catamaran, and it's often raced in fast electric events. The hull and deck are molded of very thin plastic, so the model is very light and ideally suited to spirited racing. The downside is the reduction in hull strength, which is not a problem until the model hits a solid object!

Our club has raced this hull extensively on an indoor swimming pool; set up with eight cells and an Astro Flight motor, it's extremely fast. Its disadvantage only appears when one hits the side of the pool at a high speed; the hull is almost sure to split, particularly at the speeds at which we run them. A heavier hull would be more durable, but it would also be slower.

Some kits are sold with the hull parts already joined, and this is an advantage, especially for beginners. Such kits include the Kyosho Viper and the Graupner Tornado. Building time is reduced, and the hull sections are guaranteed to be properly aligned. Usually, only minor parts such as the deck and wings have to be cut out and assembled.

Other models require that the hull parts be cut out of the vacuum-formed sheet, but this isn't difficult as long as you take your time and proceed carefully.

WORKING WITH ABS

When cutting out vacuum-formed parts, it's first best to score along the cutting line lightly with a sharp hobby knife, following the line carefully. Then go over the scored line until you've cut more than halfway through the plastic. Next, carefully fold the plastic along the cut line, bending it away from the cut itself until the plastic tears. A piece of 100-grit sandpaper on a sanding block will smooth the resulting rough edge.

When it's time to join the hull halves, I use a combination of thin and thick CA glues. I'm puzzled by the desire of many modelers to use silicone glue or sealant to join the hull pieces; they aren't good choices—too flexible. Before the upper and lower hull pieces are joined, they're quite flexible and easy to twist and bend. They gain greatly in strength and rigidity when they're firmly and permanently joined with a strong CA. Silicone allows the hull halves to flex independently, and this will ultimately allow the joint between them to leak and the halves to separate eventually. Use a strong, rigid, permanent glue to join them, and you won't have to worry about the model coming apart all by itself.

Buy a sheet of 1⁄16-inch-thick plastic to use as reinforcement material. Glue a piece to the inside of the transom to act as a doubler, and glue pieces between the hull bottom and the inside of the deck; they'll stiffen the boat noticeably. Also use the plastic to build battery platforms, receiver holders and the like.

Actually, the same 1⁄16-inch-thick plastic sheet may be used to build an entire model from scratch. If you're interested in this idea, start with a simple design, perhaps from published plans for a wooden model. The model you build may be heavier than a balsa version, but it will be waterproof, and it won't require as much time or primer to finish and paint.

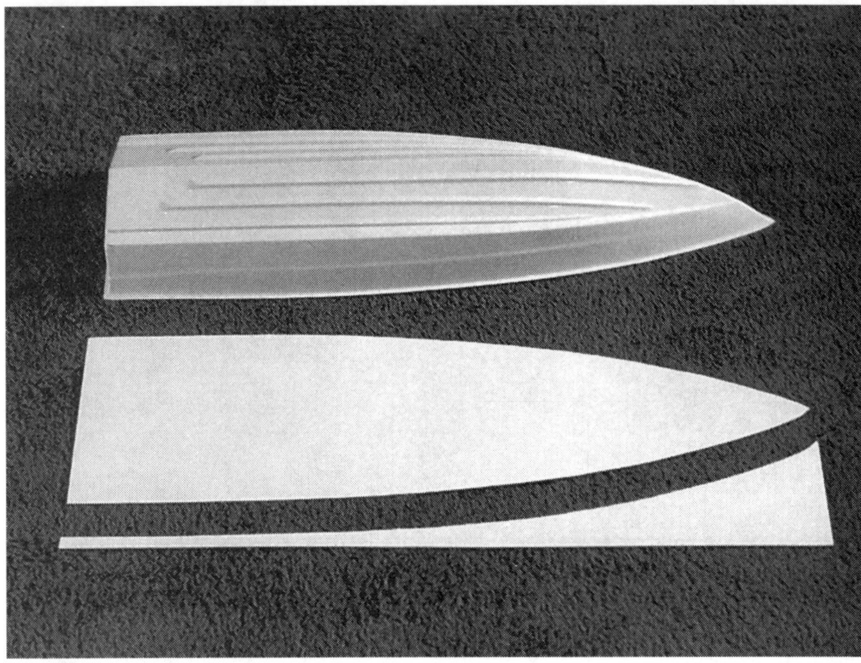

One of the easiest boats to construct is a vacuum-formed monoplane. Cut a flat deck out of sheet plastic, and glue it to the top of the hull; install the drive components and the radio, and head to the pond.

When you install the drive-shaft stuffing tube, tack it into place with a drop or two of CA, and when you're certain that it's properly aligned, mount it permanently with epoxy. Some modelers would like to use silicone glue here, apparently thinking that it would provide a better waterproof seal. As with the original joining of the hull halves, the stuffing tube can noticeably stiffen a vacuum-formed hull if it's rigidly installed. Silicone will not accomplish this and, with time, it may develop leaks owing to the hull bottom's excessive flexing.

Of course, silicone glue does have uses in a plastic model; I use it to mount servos and motor mounts (ones that have a large area to glue). It secures these items well; however, with care, they can still be pulled loose later when you want to replace or repair them.

On plastic hulls, through-hull fittings are easily mounted with silicone glow-fuel tubing. Slip a 3⁄8-inch length of medium tubing over an antenna tube, push the tube through an 1⁄8-inch hole in the deck, and push on another length of tubing under the deck to keep the antenna in place and waterproof it without a permanent installation. The same idea can be used to install rudder pushrod tubes through the transom.

I never finish painting a plastic model before running it at least a few times. I want to discover any bad handling traits or hardware problems before I invest time in painting; this allows me to correct the problems without damaging a fancy paint job.

Balsa and lite-ply are handy woods for building a light outrigger. A low parts count and mostly straight lines make this type of hydro very easy to construct.

Here, the outrigger is ready for a deck; paint it and install the hardware. The sponsons are foam and will be covered with model aircraft heat-shrink film.

I can also test that the seal around the deck is waterproof—an important consideration for all fast electric models.

Use a fine Scotch-Brite pad to take the sheen off the outside surface of an ABS boat, and then wash the surface with denatured alcohol. Don't use harsh solvents such as acetone because they'll melt the plastic! For simple paint jobs, I often use Krylon-brand spray enamels; they're inexpensive, they cover well, and they're available in many attractive colors. Of course, other paints will work well, too, and an artist can go wild with an airbrush and fancy paint jobs.

A few words of warning: to make sure that the paint you intend to use isn't of a type that will damage your plastic model, test it on a piece of scrap plastic before you start to paint the model. I once discovered this the hard way, and the mess required several hours of sanding to correct. A paint test is a small price to pay to avoid that kind of trouble.

WOODEN HULLS

Many boaters enjoy working with wood almost as much as they like running fast electric models, and I can relate to that. Though the vacuum-formed plastic boats are attractive from time and performance standpoints, there's just something about a wooden boat that turns some modelers on. In addition, some boat designs aren't available in plastic, and if the fast boater wants a model of one such hull, he'll just have to build it himself.

• **Kits.** Modelers who have no experience with wooden boats are advised to start with a kit. Examples are those offered by companies such as Dumas* and Sterling*. Most of these kits were designed long before fast electrics really came into being, so they might not be optimized for fast electric use, but they can provide the satisfaction that comes from running a model you built yourself out of wood.

The principal problem with kits like these is excessive weight, because they were designed for glow power or slower electrics. Some heavy woods can be replaced with lighter ones, but I don't recommend that you substitute lighter woods in the framing or the hull stringers; to avoid warping while they're being assembled, these areas may have to be made out of stronger, heavier wood. When you're building decking or a superstructure, heavy plywood may be safely replaced with balsa or lite-ply.

• **Scratch-building.** You may build a wooden boat from plans, such as those published by *R/C Boat Modeler*. They're well-tested, successful designs, and a modeler with some building skills should have little trouble putting together a Crackerbox, for instance. Building a boat such as this is an excellent way to learn model building, and the result should be a rewarding, successful project. Some plans have an advantage over kits in that they were

A different kind of hydro, this one uses a gear drive and model aircraft technology. It's a simple construction of balsa and lite-ply.

designed for fast electric power, i.e., they aren't older designs that have been modified to use it.

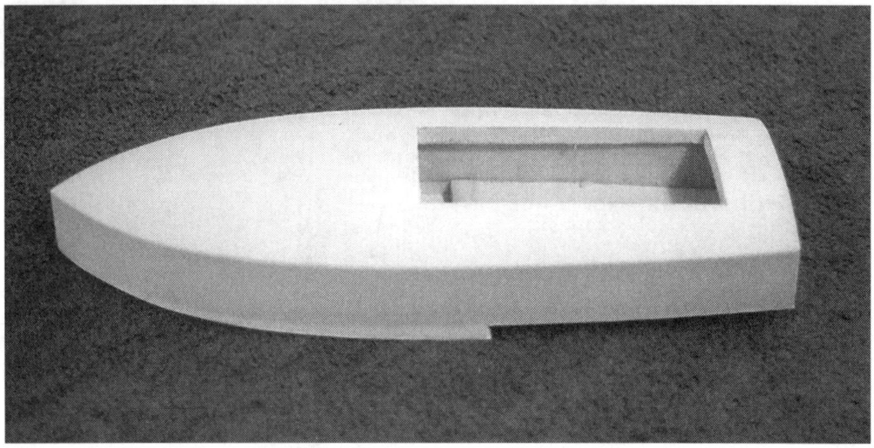

Foam is great for boat construction; it's strong and light and it floats. This model was cut out of blue Dow Corning foam (available from building-supply stores).

• **Design your own wooden model.** This does not have to be as intimidating as it sounds and, with careful planning—along with experience gained while building other wooden models—an advanced modeler should be able to design a successful boat. I've designed and built many completely or partially wooden boats, and most of them performed very well. True, a few failed to work as planned, but that happens to designers of full-size boats, too.

I enjoy building with wood because it allows me to experiment with new and different designs without making a big investment in time and money. I'll often build an unpainted prototype out of balsa to see if my ideas will actually work; if the original design is successful, I can build a fancy model later. An example is a new hydro I'm working on. It's designed for 6-cell ROAR stock racing, and I'm minimizing weight by choosing lightweight building materials and equipment. I don't know whether the idea will work, but according to my experience, I think it has a good chance of success. This model was designed for a highly specialized class, and it probably wouldn't work well in an 8-cell open motor class, but that's OK. I hope it will be a winner in the class for which it was designed; if it isn't, I'm not out a lot of time or money, because the prototype will not be built with looks or high-tech materials in mind.

BUILDING TIPS

This book is not meant to teach you how to build a model boat from scratch, but several tips are in order.

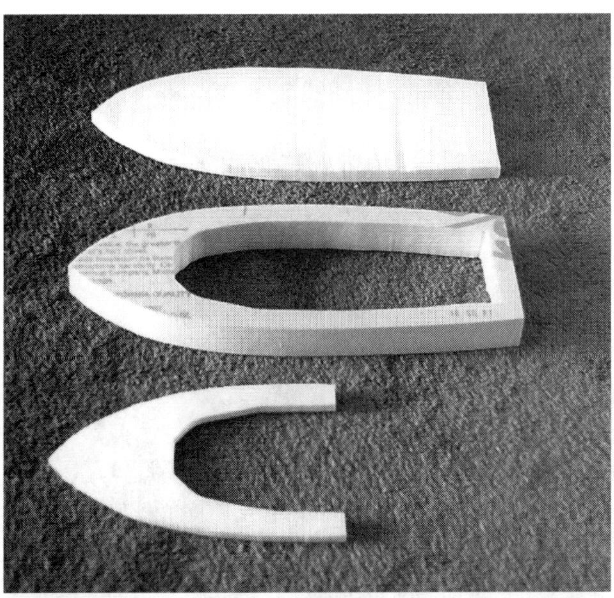

These three pieces of foam and a few scraps of sheet plastic were combined to produce a simple, but very fast, hydroplane.

• **Waterproofing.** Though it may seem obvious, wood will warp if it is allowed to get wet, and it is imperative that a wooden boat's frames and planking be completely sealed against water. Before applying the decking to a model, brush a mixture of 50:50 isopropyl alcohol/epoxy onto all the frames, stringers and planking *inside* the hull; and coat the underside of the deck before you attach it. This will help to prevent any water that gets inside (and some *will* get inside!) from damaging the wood and causing the hull to warp out of shape.

To waterproof some of the smaller pieces of wood, saturate them with thin CA. This is a great method to use on balsa, since the glue substantially hardens the soft wood. If you need to use machine screws on a wooden boat—for instance, to attach a strut to the transom—drill and tap the holes, then put a drop of thin CA into each hole. The glue will harden the wood, and you'll be able to tighten the screws without stripping the threads.

• **Trial-fitting.** This may seem elementary, but all the hardware and radio equipment should be trial-fit into place before the model is finished. When everything is in place and works, remove what you can and start to apply the model's finish.

• **Sealing.** Open wood grain must be sealed, and to keep weight down, you should use model aircraft products. I like to spread Carl Goldberg Models* Model Magic filler, thinned with water, over the hull's exterior. This soft material is easy to sand, and after smoothing down the sides and bottom, I apply a sealant such as thinned epoxy or another penetrating "paint." Then I sand this smooth and apply a conventional primer and paint. Remember to keep the weight down; if you aren't careful, a thick, heavy finish will add a lot of weight to a model.

TOOLS REQUIRED

Though fancy tools aren't required to build museum-quality models, some types will help to speed things along. In addition to the usual hobby tools, such as an X-Acto knife, sanding blocks and files, you may want to invest in a few others.

• **A small hobby sander/grinder.** This handy tool will save you lots of time when you shape parts, and its adjustable guide will ensure that square corners remain square.

• **A hand grinder,** such as the Dremel* Moto-Tool, which has a cutoff wheel, is always helpful.

• **A hand drill** (power- or hand-driven) is needed for many procedures.

• **A large drill press** is even better than a hand drill. I'd feel lost without mine, but such a large and costly accessory is not really necessary.

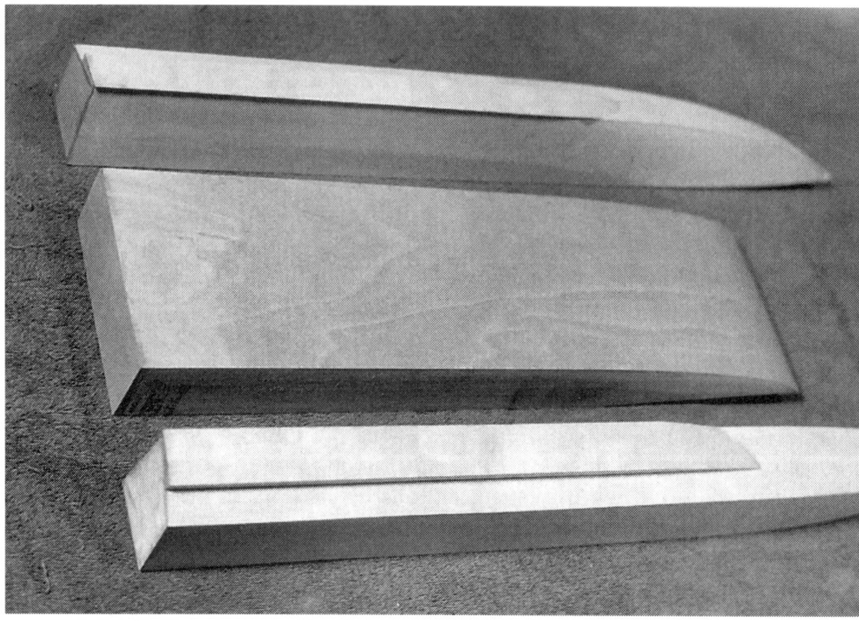

Wood and foam can be combined to produce a very strong hull. This catamaran has plywood-covered foam sponsons and a balsa center hull.

• **Sandpaper in a variety of grades**—from 80-grit down to 600-grit.

• **Rigid and flexible sanding blocks.**

• **A plastic Scotch-Brite pad** is handy for taking the gloss off plastic hulls before you paint them.

Though an airbrush would also be very useful for painting, you can obtain excellent results without one. If you can find the colors you want, cans of spray paint such as Krylon or PlastiCote will give beautiful results.

Warning: when painting indoors, be sure to use a respirator; if you paint *outdoors,* I hope you like grass and bugs in your finish!

When all the colors, decals and striping have been applied, be sure to seal everything with a couple of coats of compatible clear finish.

Wooden hulls allow the builder plenty of freedom to experiment with different designs. To increase the model's planing speed without hurting its rough-water performance, these sponson bottoms were modified by adding wedges. Pre-cut balsa strips made the change easy.

Fast electric boats need not be simple, unattractive models. This outstanding scale hydro is an exact replica of one of the first three-point unlimited racers, and it's an outstanding performer, too.

CHAPTER 10

Installations

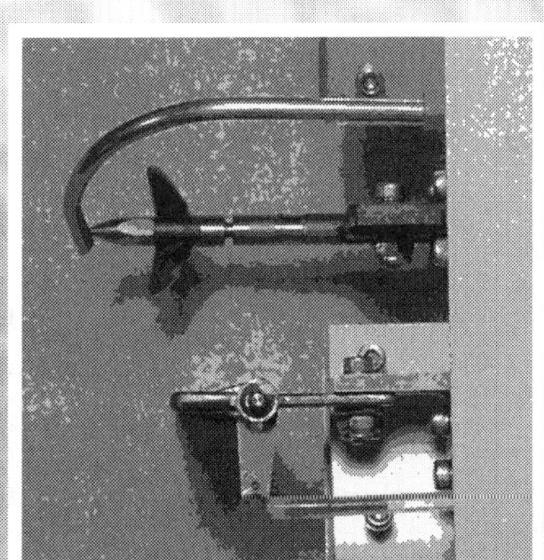

MOTORS AND DRIVE SYSTEMS

Installing the drive gear and the control gear equipment is one of the most important steps in building and fitting out a fast electric boat. A poor installation that has bad driveline/motor alignment and binding rudder linkages will cripple a boat and make it cranky, slow and unreliable. Any extra time spent on planning and installing the equipment is always time well spent. Think about what you're doing, and keep in mind the importance of making all these components work together efficiently; compromises here can return to haunt you.

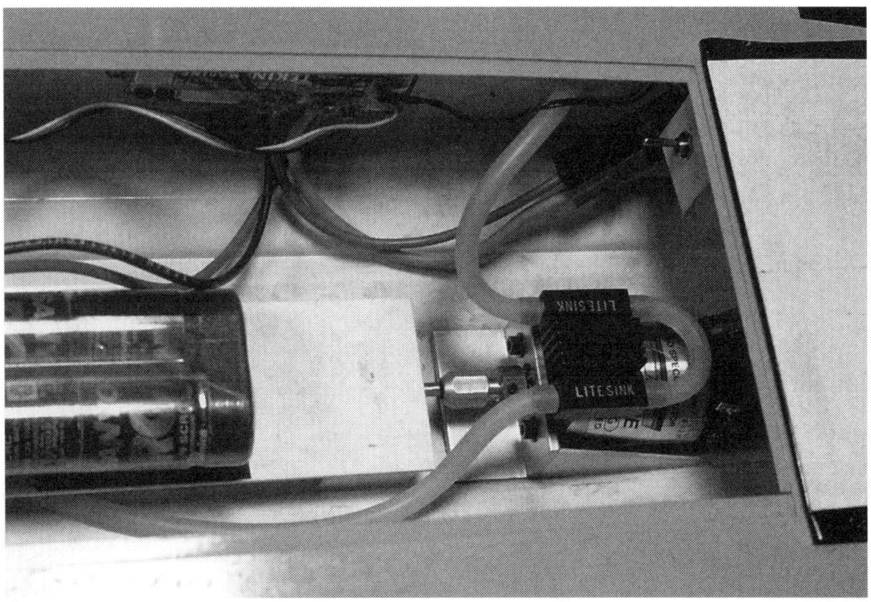

This shot of a hull interior shows a number of interesting features. The motor is firmly mounted on aluminum angle stock and fitted with a water-cooler. The ESC is mounted where bilge water won't get to it—out of the way on the side of the hull. A raised battery tray allows the pack to be moved fore and aft for optimum balance. The ESC switch is firmly mounted above the motor (don't let these switches dangle unsecured). A lip runs around the deck opening and allows the flat deck to be taped down for a waterproof seal.

• **Motor.** Most kits and boat plans include information on motor and drive-system positions, but details are often lacking. Mounting the motor is usually the first step; check the plans to determine its location. Take plenty of time when aligning the motor and the driveline; poor work here will result in poor performance later.

When possible, I like to use a silicone glue to attach the motor mount to the hull bottom. I've never had a mount come loose (until I wanted it to), and the glue's slight flexibility allows you to remove it without damaging it or the hull. There are times when epoxy, CA, or screws are a better idea, but I most often rely on the silicone.

• **Stuffing box.** After the motor has been permanently attached, line up the stuffing-box/flex-cable tube between the motor coupler and the outside strut (if used). If necessary, enlarge the hole in the bottom of the hull; you can always fill it later.

Thread the shaft/cable through the strut and the tube, and attach it to the motor coupler. Avoid having sharp bends in a flex system; just try to get the driveline parts perfectly aligned and the flex tube smoothly curved. After everything has been lined up to your satisfaction, temporarily tack the stuffing tube into place with a drop or two of CA.

The stuffing tube should be made an integral part of the hull bottom, and it should be materially stiffened in the process. For plastic hulls, cut some pieces of thin sheet plastic, and use CA to glue them inside the hull between the stuffing tube and the bottom. Use tape to cover the outside of the stuffing tube's slot in the hull bottom and to prevent any glue from running out. Mix up some slow-setting epoxy (use some phenolic microballoons to thicken the glue, if desired), and pour the glue around the stuffing tube and the hull bottom to seal and strengthen the joint. After the glue has set, peel off the tape, and sand the bottom smooth. Some additional filler may be needed to fill pinholes or voids in the hull's bottom surface.

• **Struts.** Transom-mounted struts are very easy to install. Correctly mark their positions, and drill one hole for one mounting screw in the transom. On plastic hulls, glue a doubler made out of sheet plastic to the inside of the transom to thicken and strengthen the stern. I use 6-32, plated machine screws to attach most hardware, so I drill a hole that's small enough to allow the screw to bite. I prefer to use a tap on the hole to ensure a perfect fit for the screws.

After the first hole has been drilled, mount the strut and align it vertically, then drill and tap the other holes through the strut's mounting plate. On wooden transoms, after you have drilled the holes, place a drop of thin CA in each to strengthen the wood before you tap the holes. This ensures good threads in the wood and makes the wood hard enough to endure abuse without stripping out.

BATTERIES AND WIRING

Oddly perhaps, the battery is one of the biggest advantages that fast electric boats have over their glow-powered predecessors. Glow boats are rather restricted in the placement of their drive and control equipment and, if the boat doesn't balance properly, lead weights must be used to obtain the correct center of gravity. In a fast electric boat, you can put the heavy battery almost anywhere in the hull to obtain the correct balance point; there's no need for additional weight. (This presupposes that, when you built the boat, you made allowances for adjusting the battery's position.)

I've had good luck using Velcro®-brand fasteners to hold down batteries, although some modelers prefer battery straps. Velcro® is inexpensive and easy to attach, and it allows you to adjust the battery's position easily. Plan your installation and check the balance point by temporarily placing all the boat's interior components in their respective positions. Move the battery around until the model is balanced correctly; 30 percent forward of the transom is a good place to start on most monos and tunnels. This may require a slight repositioning of some components, such as the servos and speed control.

Mark this battery position in the hull, and place the Velcro® strips in such a way that the battery can be moved an inch or two fore and aft to make small adjustments to the model's balance. You may have to build a tray out of scrap wood or plastic for the battery to sit on.

The other end of the drive shaft on the boat shown on page 64. The rudder pushrod runs straight from the servo arm to the rudder tiller. Velcro®-brand fasteners hold the battery in place. The water pick-up tube is directly behind the prop, yet it's out of the water when the boat is up on plane. The rooster-tail off the prop does a good job of filling the cooling system. The ride plates and a spin fin are attached to the transom with plated bolts.

Place the speed control close to the battery and the motor to reduce the length of wire needed to connect them, but leave enough room to allow the battery to be changed easily. To avoid electrical interference, keep the receiver as far as possible from the motor and the speed control. I often use Velcro® to mount both of these radio components.

LINKAGES

The rudder servo should be mounted as close as possible to the rudder, and there should be a straight line from the servo horn to the rudder-control arm. It's important to minimize binding to avoid having a rudder that sticks and causes control problems. I prefer to use smooth, nylon control rods as rudder pushrods, and to put small airplane clevises or ball joints at each end. Run a piece of threaded control wire though each end of the nylon tube to attach it to the hardware; the threaded rod allows you to make adjustments after everything has been installed. For a watertight seal, I glue the outer control tube to the transom. Bellows seals, such as those from Bru Line, work well, too, and they won't bind even if they're misaligned (see page 48).

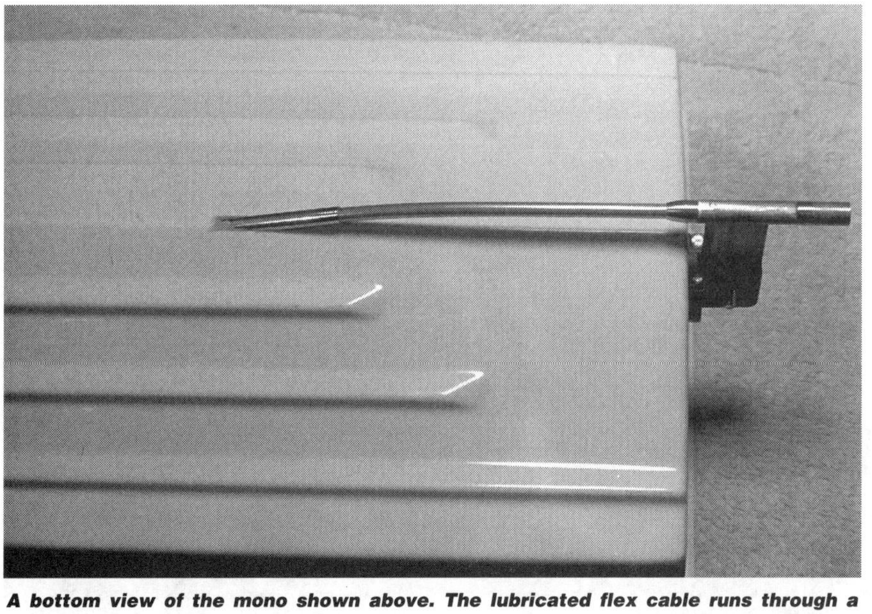

A bottom view of the mono shown above. The lubricated flex cable runs through a brass tube that enters the end of the strut nose piece. A bearing extension on the strut places the prop farther aft for improved running.

This state-of-the-art, 7-cell, unlimited racer runs a solid drive shaft that's supported in a stuffing box with bearings at each end. This is a very fast model.

KEEPING EVERYTHING DRY

One of the biggest problems faced by novice boaters is keeping their equipment dry. Many plastic boats have very poorly sealed decks, and the result is a wet interior when the boat is run in choppy water. If the boat flips (and, rest assured, a fast electric boat *will* flip) and lands on its deck, *a lot* of water can enter the hull.

A tip: if your boat flips and the deck isn't sealed well, *do not* turn the boat right-side up! Keep it upside-down and remove the hatch; this will allow all the water to drain out before the model is righted. This will keep the electrical components drier, and that will allow you to continue to race. Trust me here; it's much more fun to keep running than it is to spend time and effort on drying out your expensive equipment.

The best solution is to prevent water from entering the hull—even in the event of a flip. With some boats, this may take some imagination, but the deck can usually be sealed with tape, or an inner deck can be built to allow the interior to be completely sealed with tape. Remember, if at first your hull isn't sealed well enough to keep water out when you hold the model under water, don't run the boat until it is!

Many types of tape can be used to seal a deck, but I've had the best luck with good-quality electrical tape. It's reusable, sticks when it's damp and doesn't leave a sticky residue on your boat. Its main drawback is its appearance; used on the outside of a boat, it can look pretty tacky—especially on a scale model.

Other tapes will also work; some

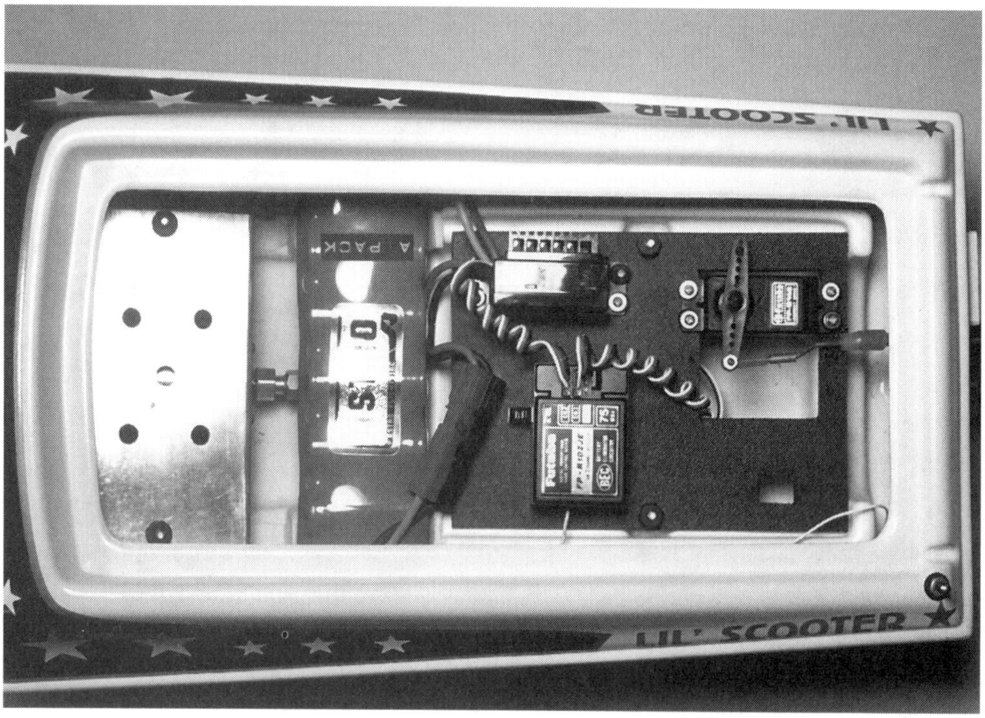

A nice, clean layout in a discontinued Parma* SkiTiger. The motor is toward the bow, under the ventilated aluminum motor mount. This model's biggest drawback is that its deck isn't waterproof.

vinyl types can be found in many different colors; even Scotch™ Magic Tape will work well but, be warned, these tapes will leave a sticky residue on your model if they're left on too long. Put it on just before a run, then remove it immediately afterward. If it's left on longer, especially if it's wet, the sticky mess that's left behind when you finally remove it may become a permanent part of your boat's finish.

I prefer to just keep the water out of my boat's interior, but with some hulls this is very difficult, if not impossible. When it's impossible, the electrical components must be individually waterproofed. Receivers can be placed in toy balloons, and servos can be sealed by applying Shoe-Goo to the wire exits and the seams; punch a hole in a piece of electrical tape and push it down over the output shaft, then re-install the servo bellcrank. ESCs pose a big problem, so the best solution is to buy a waterproof one. The motor will run if it's wet, and running it may be the best way to dry it out at the pond. It's a good idea to take along a sponge or a glow-engine fueling bulb to remove water from a flooded hull.

This Hughey twin-motor drive has excellent wiring; the wires are short and the solder points are clean. Sloppy wiring often leads to problems.

• **Flotation.** An often forgotten or overlooked installation necessity is the flotation device. Although boats without it rarely sink, I've seen a few go to the bottom without some kind of extra flotation. (All the boats that I've seen sink were recovered; but they could easily have been lost if they had gone down in deeper or murkier water.)

—*Plastic foam* is the flotation material of choice, and either the common white-beaded foam or the denser, blue, builder's foam can be used. How much do you need?

The small deck opening and the shallow compartment make this installation challenging, but with a little planning, a tidy, professional-looking job is possible. The novel motor mount uses an automotive hose clamp.

Weigh the boat just as it will be when you run it, i.e., with batteries and decking. Multiply the model's weight (in pounds) by 28. The result is the cubic inches (ci) of flotation that are needed to prevent the boat from actually sinking if all of its compartments become filled with water. For example, a 3-pound plastic boat will require 84ci of foam to prevent it from sinking. A model built out of wood will need slightly less, depending on the quantity and density of the wood it's built of. The safest route is to assume that wooden boats need as much flotation as plastic ones. Models built out of foam will, of course, need less additional flotation, if they need any extra at all.

This factory-designed installation is neat and tidy, but it suffers from the same problem as most kit boats—poor waterproofing. The radio box on the left is not waterproof and, if this type of model flips, the electronic components will be soaked. Also, the deck isn't sealed.

— *Ziploc™ bags* filled with foam peanuts or blown full of air can be stuck into open areas of a boat to provide extra flotation.

—*Balsa blocks* can also be used, but the cost of balsa is high compared with that of scrap foam.

Whichever type of flotation you choose, make sure every model has enough to prevent it from sinking. Fast electric models can and will flip; they'll hit buoys or other boats, etc. Don't be the first on your block to lose your model!

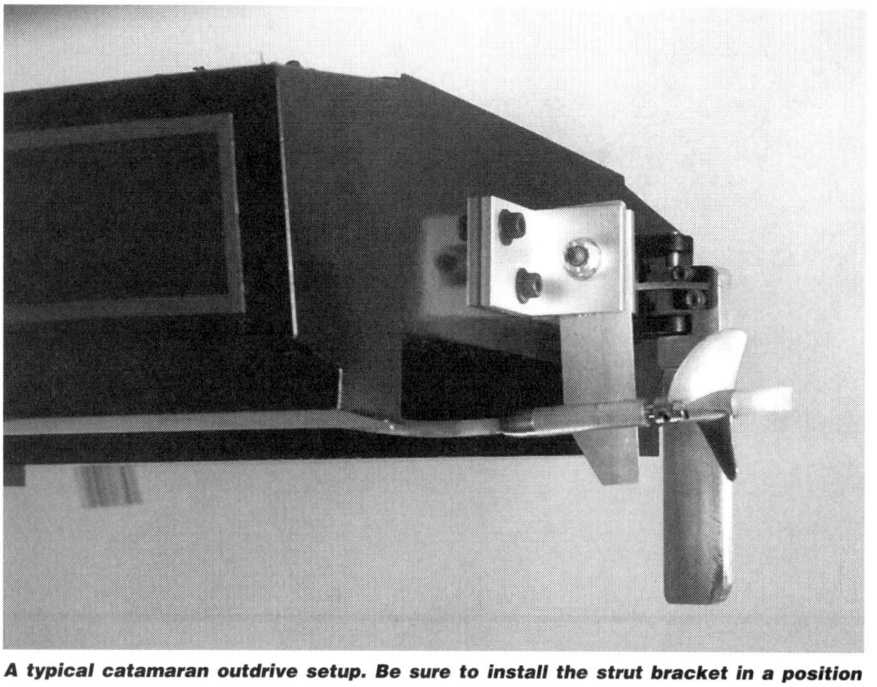

A typical catamaran outdrive setup. Be sure to install the strut bracket in a position that will allow you to adjust the strut adequately. The flex cable shown here runs through the standard Teflon® tube, but a brass tube would have provided better support. There's a gap between the strut and the drive dog; this is essential with a flex-cable drive!

CHAPTER 11

Troubleshooting

TROUBLESHOOTING

FIRST TRIP TO THE POND

When that new boat has been put together, a boater usually can't wait to get it to the pond. I understand this feeling very well, but to make that first trip successful, you should prepare at home. Of course, at home, you should assemble the model completely and try the various control systems to make certain that they work; it's a lot easier to fix a broken wire or a bad solder connection at home than it is to fix it later at the pond. There are few absolutes in life, but I will say that you *absolutely must* test your boat in a bathtub or something similar to be certain that it doesn't leak. Better to wet your model's inside at home than to lose the boat at the bottom of the pond. Don't laugh; it has happened!

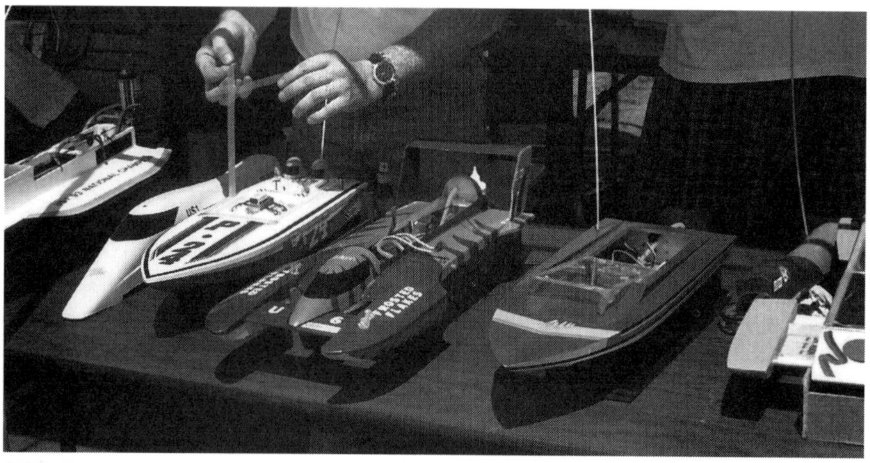

With fast electric boats, water in the electronics is a major cause of problems. This modeler tapes down the hatches on all his models.

Ideally, the initial water trials should take place on a small pond with shore access all around. At such a site, if the boat loses power or runs up onto the bank it will be easier to retrieve. If you must run on a larger body of water, try to use only one end, and keep the model close to shore at first. Always have a way to retrieve the boat if it stalls in the middle of the pond. A fishing rod and reel with a small weight or lure can be a priceless piece of equipment. On larger bodies of water, a small rowboat may be a good idea.

The first run with a new or modified model should be with a slightly smaller propeller than you intend to use later. This will be gentler on the motor and speed control, and if you misjudged the prop size, it could save you from frying your motor.

Point the boat toward the middle of the pond and give it some throttle. As soon as you are certain that you have good control, give it enough throttle to get the boat up on plane; it will draw less current there and be less likely to overheat. Be careful at first, and don't try full-throttle turns until you have a good feel for the boat. By the end of the first run, you should have an idea of whether the model is performing well; if it is not, the following troubleshooting tips should help.

OVERHEATING

Excessive current draw seems to be the biggest problem for most boaters; it overheats the motor, the speed control, or the battery. The most common causes are a prop that's too large, or a motor that's too powerful for the model.

The first move should be to try a smaller prop. Let the electrics cool down completely, then run the boat again with the smaller prop. If the small prop still overheats the motor or speed control, or if the small prop won't provide enough thrust to move the boat properly, replace the motor with a milder one.

• **Electrical system.** The experienced fast electric boater will regularly check his electrical system for excess heat. After every run, he will feel his motor and battery to find out whether they are too hot. This is a difficult test to explain to a new boater; how hot is too hot? If you can't keep your finger on the motor for at least 10 seconds, it's probably too hot. Battery heat is more difficult to judge, but if your pack is sealed in clear plastic film, look carefully at the inside of the film near the ends of the cells. If you see any moisture or fogging, the battery is venting chemicals, and if you continue to run the battery at this temperature, you'll weaken it rapidly. Obviously, if the ESC shuts down, it's too hot; be careful when you check a mechanical speed control because they can be very hot, especially if you used partial throttle.

• **Motor and battery.** You must be careful when checking for excess heat here. If you ran the boat until the battery dumped completely, expect the motor to be hotter than if you had brought the boat in with some charge left in the battery. If, after an incomplete run, the electrics are too hot, *do not* run the boat until the battery dumps without first making changes to reduce its current draw.

CONTROL PROBLEMS

Sometimes, a new boat just doesn't turn well. It may be overly sensitive and spin out, or it may not turn sharply enough. The solution here is often a simple rudder linkage adjustment. It is usually better to set up a new boat with the rudder linkage in a medium setting. That way, the steering probably won't be so sensitive that it flips the boat, but it should have enough rudder deflection to offer some control.

• **Sloppy linkage or poorly mounted components.** If the pushrod bends, it might not move the rudder enough to turn the boat. A similar loss of motion is caused by a poor servo mounting that allows the servo to move before it starts to move the linkage. Replace the weak linkage wire or rod with something stiffer, and try routing it to the rudder horn in a more direct path. Stiffen the servo mount so that there's no loss of motion.

• **Binding rudder or linkage.** These can cause steering problems by not allowing the rudder to return to its neutral position when the transmitter wheel is in neutral. This causes the model to wander after every steering command, and it can be very frustrating.

Check the rudder linkage movement very carefully with the model out of the water; you should be able to see where the problem lies. Usually, the cause is binding in the pushrod seal where the pushrod goes through the transom. Oil will sometimes solve the problem, but you might have to replace some steering components to correct it. On a high-performance model, sloppy or inconsistent steering just isn't acceptable. Spend time on correcting these problems, and your enjoyment of fast electric boating won't be compromised.

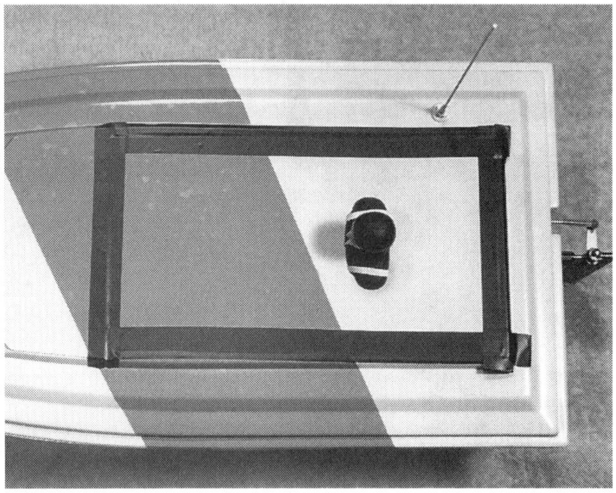

This modified Kyosho Viper has a flat deck and a tape-down hatch. The original deck can't be sealed well enough to keep out water in the case of a flip. Black electrical tape works very well to seal hatches, even when damp. If you don't like its looks, use colored electrical tape or clear vinyl tape.

MECHANICAL PROBLEMS

It's very frustrating to have a major part of your new boat break, but with high-performance models, this isn't at all rare. One of the items that most commonly breaks is the flex cable, and it usually breaks at the joint between the cable and the stub shaft. Poor soldering is one cause; either you used soft solder (always use silver solder), or you did a bad job of flowing the solder onto both parts. Another cause is rust, which weakens the cable; proper lubrication would have prevented this, and after a day's running, it's good practice to remove the drive shaft from the model and dry it.

Often, the cable simply comes out of the stub shaft, but it sometimes breaks into two about ¼ inch in front of the shaft. This means that it was overheated when soldered, or the cable was run with a sharp angle just in front of the strut. Learn how to solder well, or enlist the help of a friend or hobby shop to solder your cable. To put less strain on the spinning cable and lessen the chance of failure, change the angle of the cable where it enters the strut to make it a smooth curve. It's also probably a good idea to replace the cable at least every season.

Built according to the plans, the Crackerbox is fairly trouble-free. It uses mild motors and six cells, so overheating is seldom a problem. Its tape-down deck generally keeps it dry inside—a good thing, because in rough water, it's prone to flipping over.

PREVENTIVE MAINTENANCE

This book often refers to the need for proper maintenance of your fast electric model, and I want to repeat that warning here. A truly fast electric boat is a high-performance piece of equipment—one that works well because it's a combination of carefully selected components operating in harmony. If any of these components begin to perform less efficiently, the overall performance of the model suffers.

I've seen many modelers with beautifully constructed boats begin the season in first place, blowing the competition away with their speed and handling. As the season progresses, however, their performance drops, and the slower boats begin to catch them. The next step is a failure to

I designed this catamaran several years ago, using plywood-covered foam for the sponsons. Even if the entire hull fills with water, this model will not sink. If I built it today, I'd include a sealed motor compartment and a tape-down hatch instead of the air-cooled motor and battery compartment (see Chapter 14).

finish owing to broken parts, and the modeler may eventually even give up the hobby, disappointed with his boat and the way it let him down.

The usual cause of this is an almost total absence of boat maintenance. In a recent incident, a fellow racer flipped his boat, and because the deck wasn't sealed well, the model filled with water. A wet radio ended his racing for the night, and when he got home, he put the model away until right before the next race, when he charged the batteries and tried to get the boat to run. In addition to having damaged electrical components, he discovered that some of the bearings had rusted and seized.

Suggested regimen:
- Disassemble the boat's drive system after each day of racing.
- Dry and clean everything that got wet, and oil the components that need it, e.g., shafts and bearings.
- Remove the brushes from the motor, check them and the comm for wear and damage, and replace and repair them now, not the day before the next race.
- If the radio got wet, open the receiver and servo cases and use a blow-dryer to dry them out completely.
- Look at the outside of the model. If you hit anything out on the course, buoys and other boats included, check the hull for damage such as cracks and holes. Often, a turn fin or a ride plate will have been bent, and now is the time to find this out—not at the next race. A bent or dinged prop can easily be overlooked. A bent or damaged rudder and linkage is also a potential disaster if not caught soon enough to fix.

Keeping the model in top shape ensures that it will perform as it should, and that's the result we all want. The mindset of the high-performance modeler should be one that accepts that preventive maintenance will mean a faster, more reliable boat. If a modeler does not accept this, he is doomed to eventual disappointment.

CHECKING PERFORMANCE

A stopwatch can be invaluable when measuring the performance of your fast electric model. By setting up a course at the pond and timing a set number of laps, you'll develop a performance benchmark for that model on that day. Any speed changes that result from modifications you make, such as trying a different prop or motor, will be immediately noticeable; a longer lap time means the modification was a failure, etc.

Another important measure of performance is the current draw of your model. This is measured by installing a fully charged battery and running the model around the course at full speed until the battery dumps. Write this time down for future reference. If a well-set-up fast electric boat is to be raced in competition, running good 1400 SCR

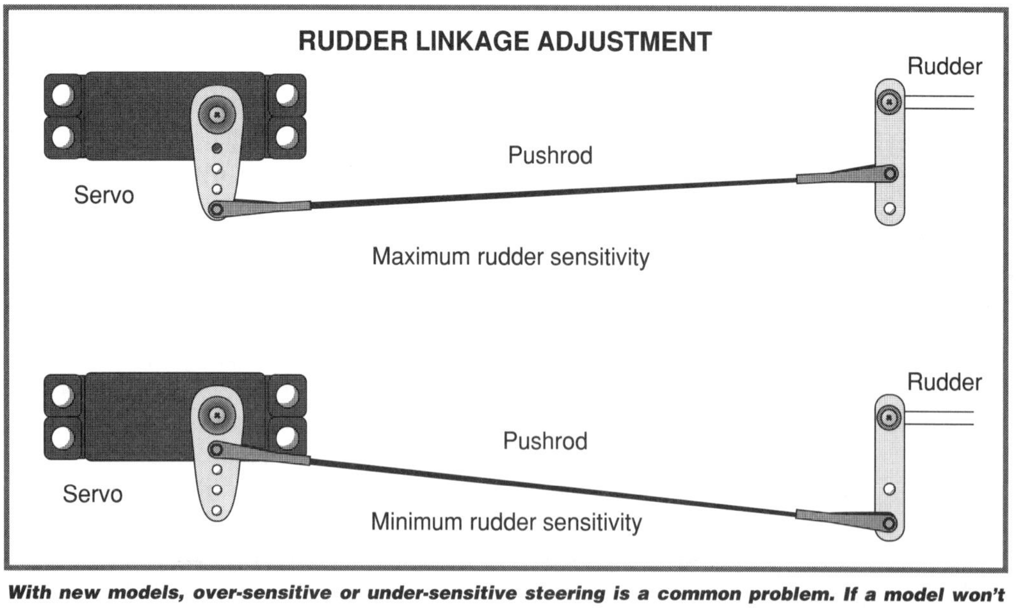

With new models, over-sensitive or under-sensitive steering is a common problem. If a model won't turn sharply enough, move the rudder linkage on the servo to the outermost hole in the servo arm and to the innermost hole on the rudder's tiller arm. If the model is too "squirrelly" and over-sensitive, move the pushrod to the innermost hole on the servo and the outermost hole on the rudder's tiller arm. If your model lacks these adjustments, it might be a good idea to modify it so that you can make them.

This is what happens if you don't include enough flotation in your model! Only the air trapped in the sponson tips is keeping this boat from a trip to Davy Jones's locker. Don't think that this can't happen to you.

cells, this time should be about 2 minutes. This means that your motor/prop/hull combination draws about 40 amps, which is about as much as most motors will stand for repeated use. A 40A current draw with seven cells produces more than 300 watts—pretty high power for a 2-foot-long boat.

I've noticed a problem in recent years with very fast racing boats running high current draws; even though the boat runs fast and well and has enough battery left to finish a race, the motor may be mysteriously damaged. Astro Flight motors had a reputation for being virtually bulletproof; they were very difficult to damage by overheating. As Astro reduced the number of winds in their newer racing motors, the power went up substantially, but efficiency dropped slightly. The result is an increase in the amount of heat produced, and—make no mistake—too much heat can destroy a motor.

In the case of the new motors, Astro Flight warns that their four-turn motor should not be run for more than about 1.5 minutes, and it should be cold at the start of the run. If you race an Astro motor and are still running after 90 seconds, chances are you're in last place anyway. Running past 90 seconds will cause too much heat to be built up in the motor and may damage it. Damage can be avoided and run time extended by water-cooling the motor.

Of course, this problem is not restricted to the Astro Flight motors. I mention them because their motors were not noted for this problem until recent years, and because the company has recognized it and has issued instructions on how to avoid it. Car motors will overheat and blow up, too, and they will seldom last as long as the Astro motors unless they're water-cooled. Overheating is *not* a motor problem; it's a fast electric boater's problem, and he is the only one who can prevent it. Either reduce the current draw to slow the buildup of heat, or don't run the boat that last little bit until the battery dumps completely.

How do motors fail when they overheat? When the terminal gets hot enough to melt solder, the motor leads often melt off. The electrical insulating coat on the armature wire can burn, and this causes shorting and additional problems. I have seen the plastic endbells on R/C car motors and the brush hoods on Astro motors soften and melt because of excess motor heat. The most common failure mode for the Astros seems to be the separation of the commutator plates from the underlying material. The motor will stop suddenly, and on opening the can, the boater is confronted with a commutator that has opened up like a flower—rather pretty, but expensive.

This large mono requires lots of power to keep up with smaller boats. It's tempting to use a large prop to push the model faster, but with direct-drive systems, this can overload the motor. Instead of abusing your motor and speed control, consider using a gear drive. That was the solution used in this model, and it works very well.

If your boat runs for more than 3 minutes, you can expect some speed improvement by increasing the current draw. The two easiest ways to do this are by trying a larger prop or one with a higher pitch, or by using a more powerful motor. If you were running a 16-turn car motor with an Octura X432 prop, try using an X435 instead. Another change would be to use a 14-turn motor with the original X432 prop. Either change should result in a decrease in run time, and it might also result in an increase in speed; time the lap speed and total run time to find out.

If you're thinking that tuning a fast electric for maximum speed can be time-consuming, you're right—especially if you want to compete. For sport running, I wouldn't go to all the trouble of setting up a course just to measure lap speed. Timing the run time is a good idea for all electric boaters, and it's easy to do because it requires only a stopwatch.

MAXIMIZING PERFORMANCE

A boater can do a couple of small, but often overlooked, things to improve his model's performance. A commonsense approach to analyzing the entire model is needed, but almost anyone can achieve positive results if he puts his mind to it. The actual speed increases may not be great, but in a close race, they might make the difference between winning and losing.

Assuming that you have the optimum motor and prop for your appli-

TROUBLESHOOTING CHART

Problem	Cause	Solution
Motor overheats	Prop too large	Try smaller prop
	Motor too powerful	Try motor with more turns
	Binding in drive	Check alignment
	Too much advance	Back off timing
Speed control hot	Current too high	Try higher rated control
		Try smaller prop
		Try fewer cells
Prop cavitates	Wrong prop	Try X Series prop
	Prop too high	Lower prop
	Hull design	Try toss launch
Battery too hot	Prop too big	Try smaller prop
	Motor too powerful	Try motor with more turns
	Bad cells	Check and replace
Connections melt	Current too high	Try motor with more turns
		Try smaller prop
	Poor connector	Try better quality connector
	Poor solder job	Learn to solder good joint
Boat won't plane	Too heavy	Reduce weight of model
	Wrong prop	Try different props
	Poor adjustment	Adjust prop depth/angle
	Wrong motor	Try motor with fewer turns
	Not enough power	Try motor with more cells
	Design problem	Try toss-launch
Poor run time	High current draw	Try smaller prop
		Try motor with more turns
	Poor boat trim	Adjust ride attitude
	High performance	Adjust thinking
Model stops suddenly	Wire melted	Replace wire/connector
	Brush stuck	Replace motor brushes/springs
	ESC shut down	Let ESC cool off
	Driveline parted	Repair
	Prop fell off	Replace prop

cation, check for anything that may waste your limited power. Friction in the driveline is one often overlooked cause of poor performance, and in almost any application, it can be reduced. Make certain that the drive shaft is perfectly aligned with the motor and the motor coupling.

Misalignment causes a side load on all bearing points, and in addition to slowing you down, this can cause an expensive drive train to wear prematurely. Fix major misalignments by moving the motor. Elongating the motor mount holes is an easy way to correct this.

Check that the turn fins and spin fins are correctly aligned with the model's center line. If they aren't, dialing in rudder trim to correct for this will cause drag that will slow down the model.

Proper lubrication is another often overlooked area. Motor bearings should be oiled after every run or two, and solid drive shafts should be removed after each run and re-oiled with a light machine oil. Flex shafts can be greased with a waterproof grease, or for reduced drag, a medium-weight (WT) oil such as 60WT gear oil can be used. Don't think that the Teflon cable tube in which your flex cable runs doesn't need lubrication; it does! Oil that cable between runs.

Details on more involved changes are included in the next chapter ("Advanced Tuning Tips"), but the information given here should get you started in the right direction.

CHAPTER 12

Advanced Tuning Tips

IDEAS FOR THE ADVENTUROUS

Some of the ideas and suggestions in this chapter are not for everyone. Many of them will be useful for most fast electric boaters, others will compromise a modeler's potential for enjoyment by complicating the model's construction or operation. Read the chapter with this in mind, and don't adopt all of these ideas without thinking about the problems, extra costs and inconveniences they may cause.

All bushing motors should have a thrust bearing fitted between the motor can and the drive-shaft coupler. On this ROAR stock motor, a ball bearing from an R/C car is installed so that it transfers the thrust from the coupler to the motor can. When the drive shaft is pushed toward the motor, the bearing should be pinched between the coupler and the can, but there must be a little play when the drive shaft is pulled away from the motor. Don't create a problem by putting the bearing in a bind.

HULL

• **Reducing weight.** Lighter boats are almost always faster and, when building or assembling a model, you should always keep this in mind. One small weight reduction may not make much difference by itself, but combined with many others, it can shave a surprising number of ounces off a heavy hull. Weight reduction can be taken too far when the strength of the model is reduced too much, but useful reductions are possible. The lighter a model is to begin with, the more difficult it will be to reduce its weight.

One area where it's easy to eliminate a few ounces is a model's removable deck. Wooden models often have rather elaborate superstructures and, for maximum weight reduction, these can be replaced with flat decks made out of balsa or aircraft-grade lite-ply. Some plastic models have heavy decks, too; the Kyosho Hurricane's 4-ounce deck comes to mind. A simple, flat, wooden deck cuts the weight substantially, but it also lacks the appeal of the attractive Kyosho scale deck.

With some hulls, inside formers made out of wood or plastic can be removed or modified in size to cut weight. To stiffen a vacuum-formed model, ½-inch-wide strips of 1/16-inch sheet plastic can be glued between the inside of the deck and the bottom to stiffen the hull with almost no weight gain.

You can decrease the weight of wooden hulls by building them out of lighter materials. Substitute lite-ply for thicker material, especially when framing. Balsa is a natural for weight reduction, but be careful because, in some applications, its use can also compromise strength too much. Ask the airplane building experts at the hobby shop questions to get ideas on how to cut ounces off your model by using model aircraft materials and techniques. Just remember, if your boat is too light and fragile, it may not withstand any impact at the higher speeds it will be able to attain. A slightly heavier but stronger model may finish a race after hitting a buoy, while a faster but ultralight one may not. You can't win if you don't finish.

Hull modifications can result in better running or turning performance if they're done correctly.

• **Sponsons.** In the search for higher speeds, I often make some changes to the bottoms of hydro sponsons. One mandatory modification is the fitting of sharp-edged ride plates. I usually use thin, stiff plastic sheet that I cut slightly larger than the sponson bottom. I then glue it into place, allowing the plate to overlap the edge of the sponson slightly. The sharp edge lets water fly free of the hull without excess drag, and it also helps the boat to grip the water in turns (see page 79).

• **Ride surface.** Another modification is to flatten the ride surface. Some model hydros have a sponson bottom with a steep vee (a deadrise). This helps the model to negotiate rough water, but it can limit top speed by reducing the lift created. Often, I'll make a new sponson bottom by gluing a strip of Plastruct angle to one edge of a piece of sheet plastic; then I glue this assembly to the old sponson bottom. The result is a flattened bottom that gives more lift while providing sharp edges for improved turning.

Additional modifications include: an increase or a decrease in the sponson's angle of attack; the addition of small steps; and the sharpening of the chines and transom edge on monoplanes. These can be made out of sheet plastic and plastic putty.

• **Drag reduction.** Underwater drag is one of the most damaging types of drag, and any reduction here is very good. Modelers who run for speed records will remove everything they possibly can that sticks down below the surface, and they will reduce the size of items that they can't remove. Spin fins and turn fins will be taken off, water-cooling pick-ups will be removed and the holes plugged, and the rudder will be cut down as much as possible without losing control of the model. What's left will be sharpened and polished to a high shine. Keep in mind that, for normal running, and even for oval racing, this is a bit extreme, but the message is clear; minimize underwater drag without destroying the model's turning characteristics.

One last comment on underwater drag: there is some disagreement on the best way to decrease drag on a running surface. Some boaters advocate the use of a good wax, others want to polish the bottom with 600-grit wet emery paper, and a few duplicate the full-scale practice of using a special graphite paint. I've tried wax and the 600-grit polish, and I didn't see any difference, so at present, I'll keep using a good automotive wax. That may change if I find an advantage to using another method.

This worn motor should be rebuilt, and the armature is the major component in need of repair. The commutator (comm) on the right should be turned on a lathe to remove the grooves and burns on it. While you have the armature out of the can, check its balance and correct it if necessary. Find the heaviest pole, and drill small holes in it to decrease its weight to that of the other poles. Don't drill too deeply, because you might weaken the stack's laminations and damage the armature beyond repair.

DRIVE SYSTEMS

Maximizing performance means minimizing parasitic drag and friction wherever possible. When you install the drive system, take extra care to ensure perfect alignment between the motor, coupler, shaft, stuffing box and strut. Any misalignment will waste power that could be going to the propeller to drive the model faster.

Some kits come with a solid coupler between the motor and the solid drive shaft; this works well if the alignment is perfect, but if the hull flexes even a little, the momentary misalignment will bind the shaft and slow it down. Replacing this cheap coupler with a high-quality universal joint, such as one of those offered by Graupner, will help to eliminate this source of power loss (see photo on facing page).

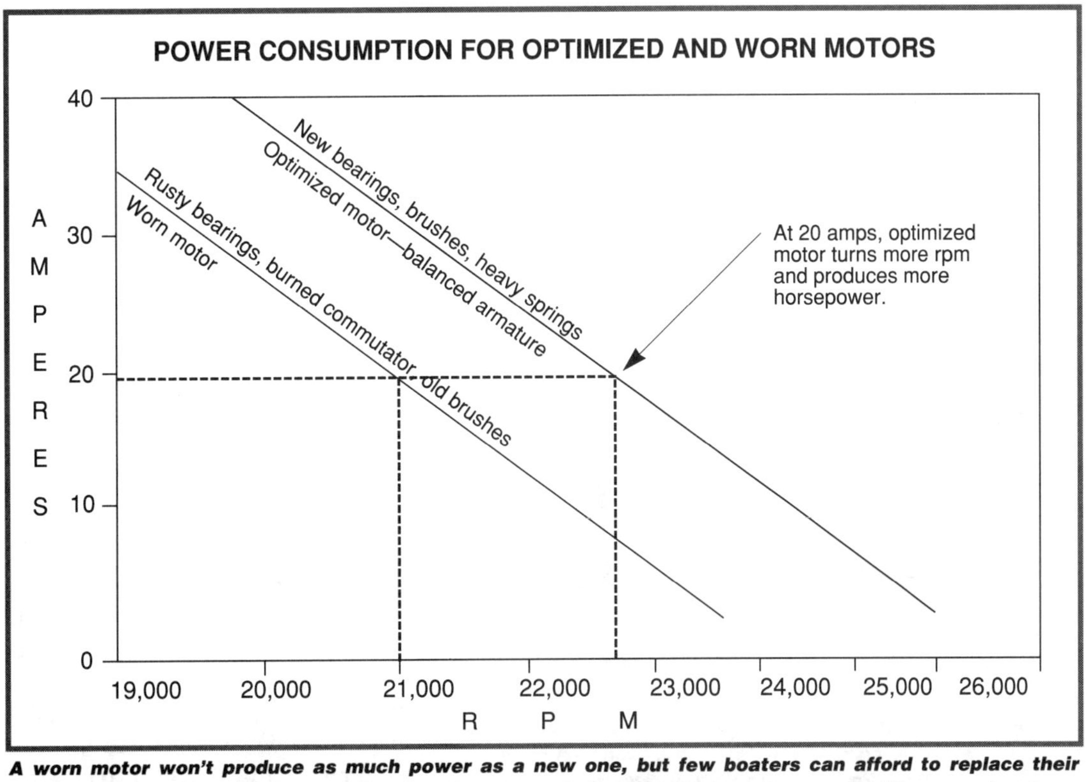

A worn motor won't produce as much power as a new one, but few boaters can afford to replace their motors whenever they wear out. Turning the comm, balancing the armature and adding new bearings, brushes and springs will restore your old motor for a fraction of the cost of a new one.

Most vacuum-formed kit boats have rounded sponson edges like those shown here. Ideally, all the edges that come into contact with the water should be sharp enough to allow the water to flow cleanly over them with minimum drag. This hull could be run as is, but its performance would suffer.

Most fast electric models are set up so that the motor bearings transfer all the driving force from the propeller to the model. Good-quality ball bearings in the motor endbell can do this without problems, but motors with bushings will overheat and slow down when forced to take thrust loads as well as rotational loads. The solution is to install a thrust bearing between the motor housing and the drive-shaft coupler. One-eighth-inch thrust bearings from an R/C car will work well; just be sure to lubricate them before every run. Set the bearing up to allow some end play between the motor housing and the coupling, or the bearing may not function properly. Although it probably isn't necessary, a thrust bearing can also be used with motors that have ball bearings.

• **Reducing friction.** Solid drive shafts that run in brass stuffing boxes have a lot of unwanted friction, and anything that's done to reduce this will help to speed up the model. One easy way to reduce friction is to lightly coat the shaft with a fine polishing compound, install it in the stuffing box, and run it at a low speed for a minute or so. This laps-in the shaft and the stuffing box, and that takes off the rough spots and makes the entire tube a smooth bearing. Be sure to clean off all the polishing compound, and lubricate the shaft with a light oil before every run.

To reduce friction further with a solid drive shaft, use a stuffing box that's larger inside than the shaft's diameter. Brass, aluminum, or even graphite arrow shafts can be used for the tube; put bearings at each end, and friction will decrease substantially. The bearings need not be too elaborate, and 1-inch-long pieces of brass tube at each end of the stuffing box will work as well as bronze bushings if they're lubricated regularly. An even better idea would be to use Teflon® or bronze bushings that are intended for drive struts. Some boaters will even install ball bearings, but these can require frequent replacement because they're prone to rusting when they're run under water.

POWER COMPARISON CHART

Kit System		Optimized System
Welded tabs	BATTERY	Soldered copper bars
18-gauge plastic	WIRE	13-gauge silicone
Mechanical resistor	SPEED CONTROL	High-frequency ESC
Cheap plastic/steel	CONNECTORS	Low-loss plated
3.9V loss	RESULTS @ 30 AMPS	2V loss

Comparison of power delivered to the motor by kit system and optimized electrical system.

A few minor changes to the electrical system can result in dramatic power increases. Here, a typical kit boat's voltage losses are compared with an optimized system in the same boat. The optimized version supplies almost 2 more volts for the motor than the factory kit does, and that means that the boat goes faster!

• **Reducing drag.** Flex drives are easier to set up because, to some degree, they are self-aligning. Still, taking care while installing them can result in lower drag later. Most modelers run the flex cable through a Teflon® liner, but many forget that they need to lubricate the cable, too. A waterproof grease such as that offered by Prather* works very well and will stay on for several runs. The grease, however, is thick and will create extra drag on the cable; substitute 30W motor oil or 60W gear oil to get extra power to the prop.

I prefer to run the flex cable through an ⅛-inch-i.d. brass tube rather than the Teflon® one that's usually installed. If the unsupported tube between the hull bottom and the strut is more than 3 inches long, a stiff brass tube will support the drive cable much better than the highly flexible Teflon tubing will. This results in more effective thrust transfer to the motor and a lower loss of thrust through cable deflection. I have used this method for years without wear problems, but I am very careful to keep the cable well-lubricated with grease or oil.

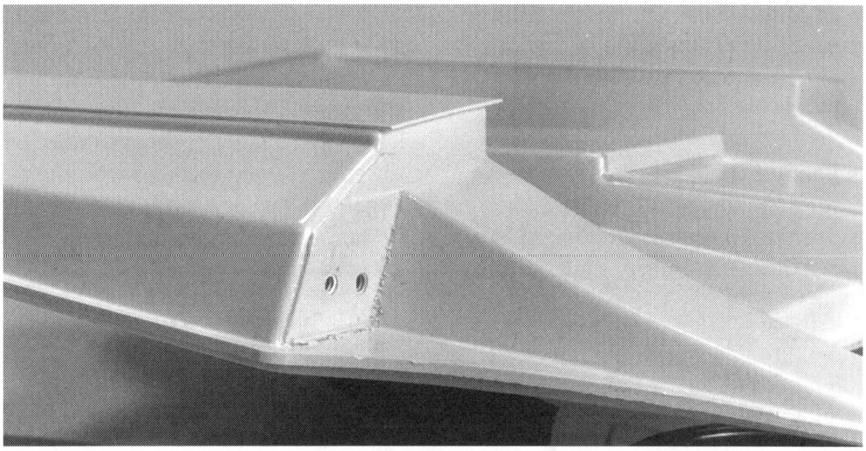

Ride plates have been installed here to sharpen the sponson edges. This results in a faster boat owing to a cleaner exit from the water. With the sharp edges, turning is also improved. An extra piece of sheet plastic was used as a doubler for the turn-fin mount. It's a good idea to use doublers wherever a strut, a rudder, or a fin is attached to a plastic hull. Doublers strengthen the hull and help to prevent it from being torn when the fin or the rudder hits something.

• **Gear clearance.** Gear drive systems can improve your boat's performance significantly, but they also increase its complexity. A common mistake made by novice boaters occurs when they set the gear clearance—or lash—between the two gears' teeth. The usual error is to set the clearance too small; this results in binding and an unnecessary power loss right at the gears themselves. Although the clearance on some gear drives can't be adjusted if you can adjust the clearance, the accepted value is usually 0.004 inch. I use a slip of typewriter paper to check clearance; if the paper can be rotated between the gears with a snug fit but without being torn, the adjustment is good.

• **Maintenance.** This is another necessity with gear drives, but it's often overlooked. Gears should be kept clean and replaced if they're obviously worn. Bearings must be very smooth and quiet; if they aren't, you'll lose some of the gear drive's power advantages to excess friction. The drive must be mounted securely on the hull in nearly perfect alignment with the drive shaft/flex cable. Set up and maintained correctly, your gear drive will be an asset, instead of a liability.

• **Propwalk.** Fast electric models that run surface-drive props will, to some extent, suffer from propwalk, which causes the model to turn slightly to the right with the rudder centered. The usual correction for this is to dial in a little left rudder trim to make the boat track straight ahead, but rudder deflection causes drag. It's better to offset the drive-thrust angle slightly. Modify the strut mount or the stuffing box to allow a 1- to 3-degree change to the left in the thrust angle; this should correct the model's tendency to drift to the right without introducing that drag-producing rudder trim.

BATTERIES AND WIRING

Anything electricity passes through causes some power loss owing to resistance, and the result is that a lower voltage is delivered to the motor from the battery. By reducing these voltage losses, we can supply additional power to the motor and make the boat go faster. One aspect of resistance is that the greater the current flow is in amps, the higher the voltage loss is. An example is wiring: 10 inches of cheap wire might have a resistance of 0.015 ohm. If your model is cruising along rather slowly and drawing only 15 amps, the voltage loss from this wire will be 0.2

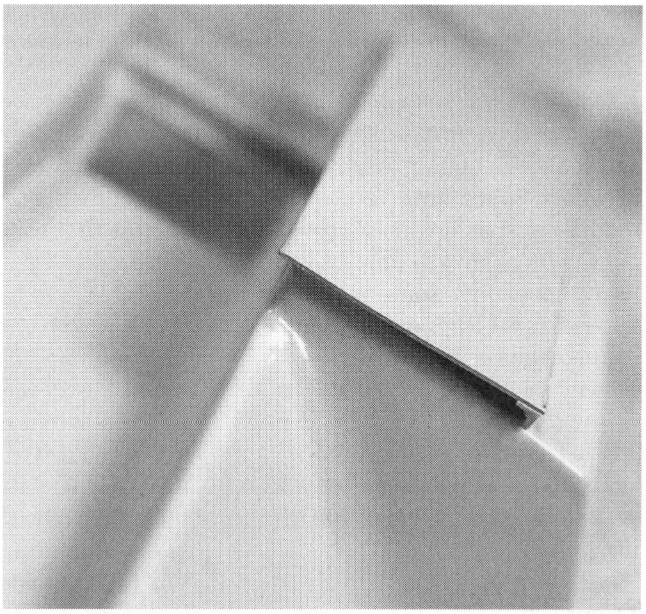

Here, a piece of sheet plastic and plastic angle stock were used to both flatten the sponson bottom and provide sharp edges for cleaner running and better turning. This modification resulted in a dramatic improvement in the model's performance.

volt. If, in a race, you increase the current to 40 amps, the voltage loss will be 0.6 volt—a substantial amount.

• **Use heavier wire.** This should be your first change; 12- to 14-gauge silicone-covered wire has very low resistance even at very high current flows. Plan your installation carefully, and use the minimum length of wire necessary. To minimize resistance-induced loss, hard-wire all components permanently into place with neat solder joints. Some boaters use fewer than 8 inches of wire to connect their motor, ESC and battery; that's not much wire! Hard-wiring requires the use of a soldering gun to replace or maintain the electrical components, but the voltage available for the motor will be maximized.

• **Connectors.** Next in line for change are the connectors. Because they have very high resistance and wear out quickly, standard Tamiya-type connectors with steel contacts aren't too good for fast electric boats. High-quality connectors by Sermos, Litespeed and Astro Flight, among others, have less than a quarter of the resistance of the cheap connectors and are almost as good as no connectors at all.

Standard, assembled battery packs are usually put together with small welded tabs between the cells. These are OK for general sport use, but not for high-performance boats because they have a very high resistance. I discovered this years ago when I noticed that the plastic shrink-wrap had melted where it touched a tab. High heat usually means high resistance, and the solution here is to use packs that are assembled with copper straps or braided wire; if soldered correctly, these have much less resistance. If you wish to make your own packs, be sure to use a soldering iron that's large enough to melt the solder adequately, and use a solder and liquid flux designed for electrical applications.

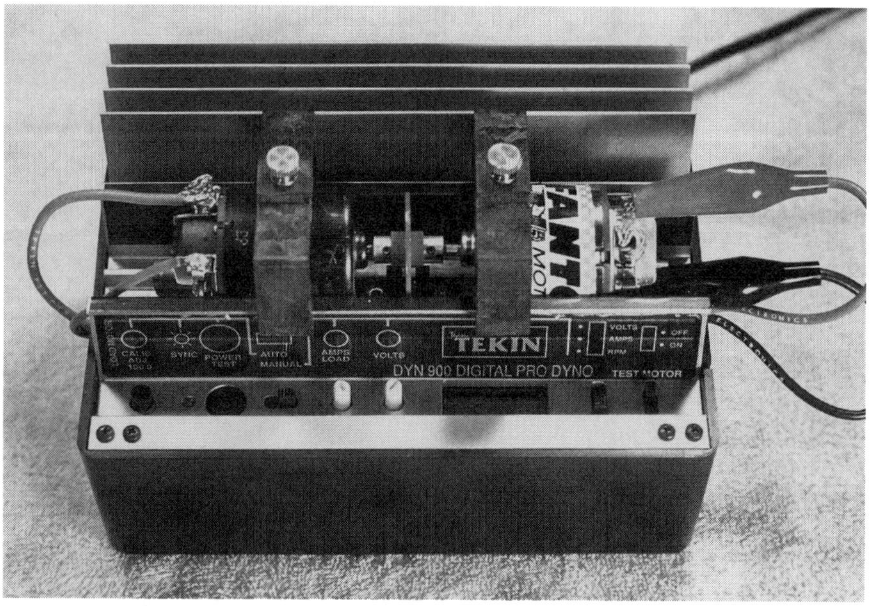

To maximize performance, advanced modelers can tune their motors with a motor dyno such as this Tekin model. Dynos aren't necessary for top performance, but for those who enjoy the more technical side of R/C boating, they extend the tuning to another level. Don't run out and buy a dyno unless you really understand electric motors and their design. Most boaters are served well by making changes to their motors and then running them in their models to check for improved performance.

• **Battery cycling.** Some top-level racers insist that, to maximize performance, even SCR cells should not be cycled more than once a day. Other racers run only matched packs, and they re-match their packs monthly using exotic battery-matching equipment. Still others sell off the packs that they've cycled more than 50 times, regardless of how well they still perform. I can't say that these techniques are necessary; in fact, I feel they are excessive. Still, the minor advantages they may provide are important to some racers, and if you lose heats by only a second or so, exotic battery practices may help you.

• **Electronic speed controls.** The ESC is an integral part of the wiring, and for fast electric use, a high-performance ESC must be chosen. Look for a unit that will handle high current draws and one with a low "on" resistance. The higher the current draw and the lower the resistance, the better. Good-quality ESCs designed for R/C cars will work well in most marine applications, but a special marine ESC is a good investment.

Some of the best ESCs, such as those from Hughey and Astro Flight, do not have a BEC supply for the radio system. Also, if your boat's motor/prop combination draws very high amps under initial acceleration, it may depress the voltage available to the radio so much that you temporarily lose control of the ESC and the rudder. In these instances, the solution is a battery pack that's exclusively for the radio system. Because weight is your enemy, use a 4- or 5-cell pack made of tiny 50 to 80mAh cells. Because of their lower capacity, they may have to be recharged between every race heat. If you use these packs, make sure that they will last for an entire heat. An out-of-control electric rocket is not a pretty sight.

One other way to correct a momentary loss of radio power caused by drive motor current draw is to use a Novak Stutter Stopper. This is a special capacitor that stores a little electrical power as soon as you turn on the ESC. Plug it into the receiver battery socket; if the current usually supplied by the BEC drops too low, the Stutter Stopper will feed its own supply of power to the radio system to keep it operating. This relief will be very brief, but it will often solve the problem without the cost and side effects of a separate radio battery.

To recap some of our electrical changes, let's compare a kit boat's typical system with our optimized high-performance system. A kit boat with a mechanical speed control running on seven cells and a big prop that draws 30 amps may have a system voltage loss of about 3.9 volts. Our optimized system (under the same conditions) has a loss of just 2 volts—only half that of the kit system. Another way of looking at this is that our high-performance system delivers 6.4 volts to the motor, while the kit system delivers a pitiful 4.5 volts. The changes we made were almost like adding two cells to the kit boat's battery pack!

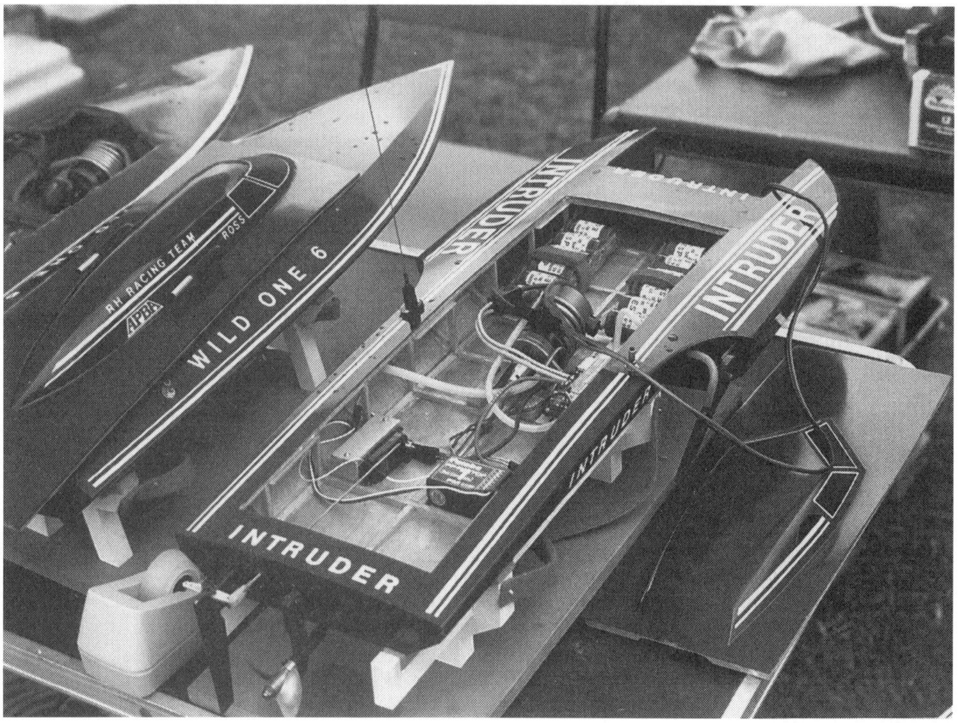

Most top racers have learned that a neat, orderly installation is necessary in high-performance boats. This model is a beautiful example: this clean, organized installation helps to eliminate a number of potential problems and shows that the modeler is an experienced, winning racer. We should all try to emulate this installation.

MOTORS

Regardless of the motor we choose to power our fast electric model, there are several things we can do to improve its performance.

- The commutator (comm) must be in good shape; if it's grooved or burned, it must be turned on a comm lathe until it's like new, or the armature must be replaced.
- The bearings must be clean, properly oiled and not rusted or damaged. Replacing them with bearings of the highest quality available will add a few hundred rpm to a high-speed motor.
- The motor's can must be in good shape and have the best magnets available; the magnets in an old can may have been weakened with use or may be of an older design.

- **Install a stator ring.** You can improve the magnetic performance of your motor by installing a "stator ring." This ring is simply a piece of iron that's wrapped around the motor's can to help keep the magnetic field focused within the motor. Some modelers make their own out of worn-out motor cans, and Trinity offers a wraparound product that does the same thing. When you increase a motor's magnetic field, you also increase its efficiency and potential power. The stronger field may actually decrease the motor's free rpm, but it will also produce more torque than it previously did. Because of this, the motor may be able to turn a large prop faster, or it may allow you to turn an even larger prop at the same speed.

Expensive, hand-wound modified motors are the cream of the crop among R/C car motors, chiefly because their makers spend extra time and money on parts and assembly. Still, for those who are willing to do the work, there is room for improvement even on these motors.

After you have disassembled the motor completely:

- Check the armature's balance on a propeller balancer. If your armature won't fit on your prop balancer, you may have to make a special armature balancer out of wood and two razor blades. Spend plenty of time on this, and make the armature balance perfectly by carefully drilling small holes in its heavy stack. An out-of-balance armature will steal power that could be used to turn the propeller.
- Correct alignment of the brush hoods is necessary to ensure proper motor timing and electrical balance. Several manufacturers make special brush-alignment tools to check the

brush hoods, but a 5⁄32-inch-drill bit will work well, too; use the full-diameter un-fluted end. If the tool won't pass freely through both hoods (from one side to the other), loosen the brush screw on one side, and move the hood until the tool will pass through both. Tighten the screw securely and reassemble the motor.

Improvement can be made to stock, tabbed-endbell motors, too. Without disassembling the motor, you can use special tools to check brush-hood alignment, and the hoods can be moved into their proper places as detailed for the modified motors above.

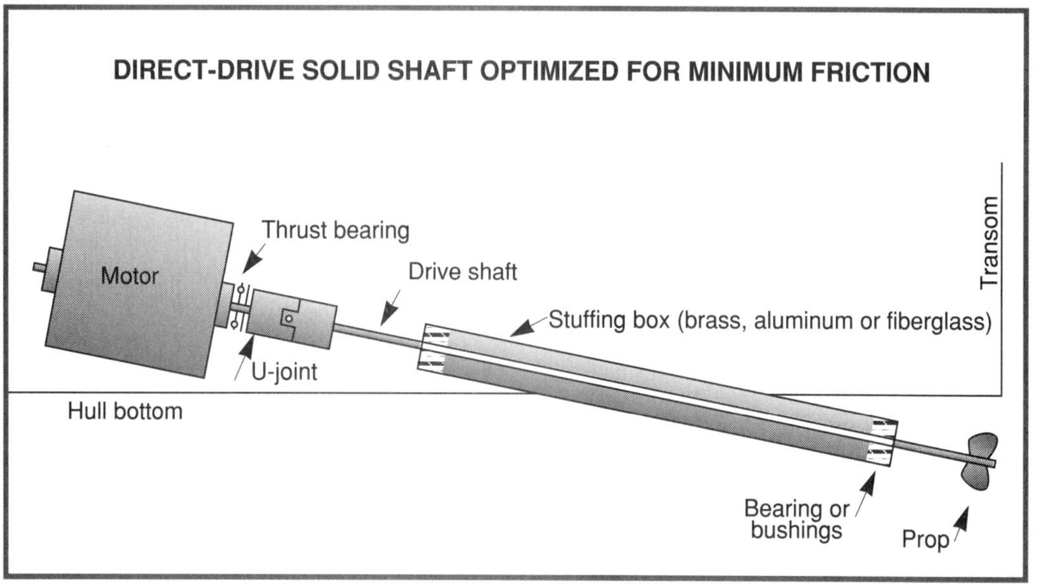

This direct-drive solid-shaft system is optimized for maximum performance. The shaft is made out of drill rod to ensure straightness, and it rides in two bearings at either end of the stuffing box. To eliminate power losses caused by poor alignment, a U-joint connects the shaft and the motor, and a ball bearing takes the thrust load off the motor's rear bearing or bushing. Although the improvements offered by each change are minor, together, they result in very noticeable performance enhancements.

• **Clean the bushings.** Another way to gain speed in stock-type motors is to lap the bushings and shaft until they fit perfectly. Mix a very fine metal polish (such as Brasso or Simichrome) with comm drops (in a polish/comm drops ratio of about 1:3) to make a lapping compound that can flow between the shaft and the bushing. Clean the motor bushings with motor cleaner to remove any oil, then apply a drop or two of the lapping compound to each bushing.

Rotate the shaft by hand, moving it back and forth in the can to work the compound between the shaft and the bushing. Then connect the motor to a low-voltage source (two to three cells is ideal) and run it slowly for about 5 minutes. If necessary, stop every minute to re-apply the compound, and take the time to work the shaft back and forth, always working fresh compound more deeply into the bushing. After 5 minutes, the compound will be darker because it contains the metal it removed, and both bushings will be well-polished. Thoroughly flush out both bushings with motor cleaner to remove all the compound, and move the shaft around by hand. (Wear rubber gloves to protect your skin from the motor cleaner.) Allow the motor cleaner to evaporate completely, then re-oil the bushings with a good-quality oil.

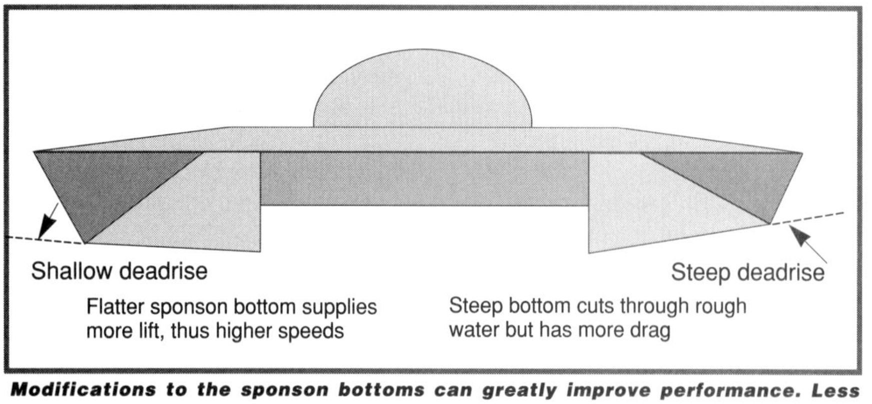

Modifications to the sponson bottoms can greatly improve performance. Less deadrise—and, thus, a flatter sponson—creates more lift, which reduces the friction between the boat and the water. If a model is too unstable and constantly blows off rough water, steepening the deadrise can reduce lift and improve rough-water performance.

• **Solder the wires.** With the basic motor in topnotch shape, solder the speed-control wires directly onto the motor terminals for minimum resistance. Soldering the brush leads to the terminals is also a good idea; screw eyes don't always make good contact. Under the high amp draw of fast electric use, hard brushes will last longer. For maximum torque and minimum arcing, use heavy (purple or black) brush springs to keep the brushes firmly in contact with the comm.

• **Choose the right brushes.** Hard

brushes are the usual choice for high-performance marine motors, chiefly because of their superior wear characteristics under high amp draws. They don't wear the comm excessively, and they conduct high current flows well. For these reasons, they are recommended for most fast electric boats.

A hard brush with a high silver content will produce noticeably higher power levels owing to its superior current conductance. The downside of this advantage is such extreme wear on the comm that you may have to re-true it after only one run! For speed-record attempts, or perhaps for national-level oval competition, these brushes can be used; but obviously, their disadvantages outweigh their advantages for any regular use. Few boaters want to rebuild their motors after each race heat.

Some record-setting racers run "disposable" motors. National racing organizations have record categories for "stock" motors, which, as I write this, are considered to be any tabbed-endbell motors with bushings. Most (but not all) motors of this sort are ROAR '91 types with 27-turn armatures. Notable exceptions are the motors by Fantom Motors*. This small company winds armatures that have between 9 and 18 turns. Obviously, these motors will produce far more power than a ROAR '91 motor, but they cost less than typical modified motors. Other motor builders also offer a limited selection of low-wind bushing motors.

These Graupner Tornadoes were built thousands of miles apart, yet their performances were very close owing to the experience and building skills of their racers. The winner gave the other boater a few tips at this national event; next year, their performances should be closer yet.

Some racers who chase national records set up their boats with a low-turn Fantom, install silver brushes and make a few runs. With batteries that dump in less than 60 seconds or so, the current draw for a pair of passes through the 1/16-mile time traps surpasses 80 amps. Combine silver brushes with a 10-turn motor and an 80A current draw, and major motor damage is almost guaranteed. The motor won't be usable for racing until the comm has been rebuilt and, if a racer doesn't have a stock-motor comm lathe, the motor often ends up being dismantled for parts. Under current rules, stock motors can't be disassembled and rebuilt.

It's a fact of life for high-performance boat motors that their parts will wear out. Although this seems obvious, it's surprising to see some racers trying to compete with motors they've run all season without having rebuilt them; sometimes they use the same brushes, too. This causes a gradual reduction in motor power; so gradual that the boater might not notice it until he starts losing heats by larger and larger margins. The brushes are wearing out, burning and increasing their resistance. The comm is getting burned and grooved, and that reduces its capacity to conduct current at high rpm. Brush springs are weakening because of the heat, and they aren't holding the brush in firm contact with the comm. Even the bearings may be corroding, and that causes the extra drag of rough balls and scratched bearing surfaces.

• **Maintain your motor.** To keep your boat running in top form, maintain your motor! Replace the brushes when they look burned, or after every six or eight heats—maximum. Brush springs should be replaced when the brushes are. Tear the motor down after each race day, and inspect the comm for damage. If it's badly burned or grooved, have it turned.

Remember that a comm can be turned only three to five times at most, depending on its condition and the skill of the lathe operator. Buying a new armature is cheaper than buying a new motor, and it's easy to replace. While the can is open, rotate the bearings with a pencil to check for roughness or noise. Replace marginal bearings with those of the best quality you can find.

This may all seem like a lot of work, but if you're really interested in top performance, you'll complete the required maintenance regularly. It's false reasoning to buy a high-priced super motor and expect winning performance, but then fail to maintain it and, consequently, see its power level fall below that of a newer, cheap, low-performance design. Following a better maintenance program could mean victory when running against a racer who has neglected his equipment.

This discussion of motor maintenance applies to the common R/C car motors. More exotic ones such as the Astro Flight cobalt racing motors will usually last much longer between brush/spring replacements, and their comms seem to wear better, too. They should still be torn down every few races so that you can check the comm and the brushes, and their bearings are just as prone to rust and damage as those in standard R/C car motors.

These tips are not the only "speed secrets" that are used by top racers, but they are some of the most useful. Obviously, some of them have no place in *sport* boating, and some are poor ideas even for racing. Always keep in mind that, although a high-performance technique may make your boat run faster, it may reduce its reliability or practicality. A boat may lose its appeal if it becomes too troublesome to run.

CHAPTER 13

Organized Racing

ORGANIZED RACING

WHY JOIN AN ORGANIZATION?

When I first started to run R/C boats (many years ago), I pretty much went at it alone and almost never ran with others. At the time, this was OK, because, as a beginner, I found the novelty of an R/C model was entertaining enough; and any mistakes I made were known only to me, and I had no need to feel embarrassed.

As time went on, though, I started to find that after completing a model and getting it to run well, I became bored with it and had to start building another one. After all, there's only so much you can do with a fast R/C boat by yourself.

That's why I say that one of my best decisions was to join a group of other boaters in organized racing. I was intimidated at first, because I wasn't too sure of my boat's ability (or my abilities) to run with other modelers. I soon found that my apprehension was unfounded; the others were just folks like me who enjoyed the hobby but wanted the challenge of competing. I soon made new friends and began to worry less about how my boats did in competition; I looked forward to the next race because I enjoyed the company of the people there.

Three 1/16-scale unlimited hydros exit a turn at the 1994 APBA Electric Nationals. Racing your model at a national event is very exciting and lots of fun, even if you don't win every time. You don't know how fast your boat is until you've raced it against the best in the country.

A move out of the area put my R/C activities on hold for several years, and when I returned to Alaska, there was no longer any organized R/C boating. Fast electric boats were then gaining in popularity, and I returned to building them. I met another boater who wanted to race, and together, we organized a new fast electric club. This club has grown steadily and now annually sponsors more than 20 races. The members aren't all serious competi-

TYPICAL ELECTRIC OVAL RACE COURSE

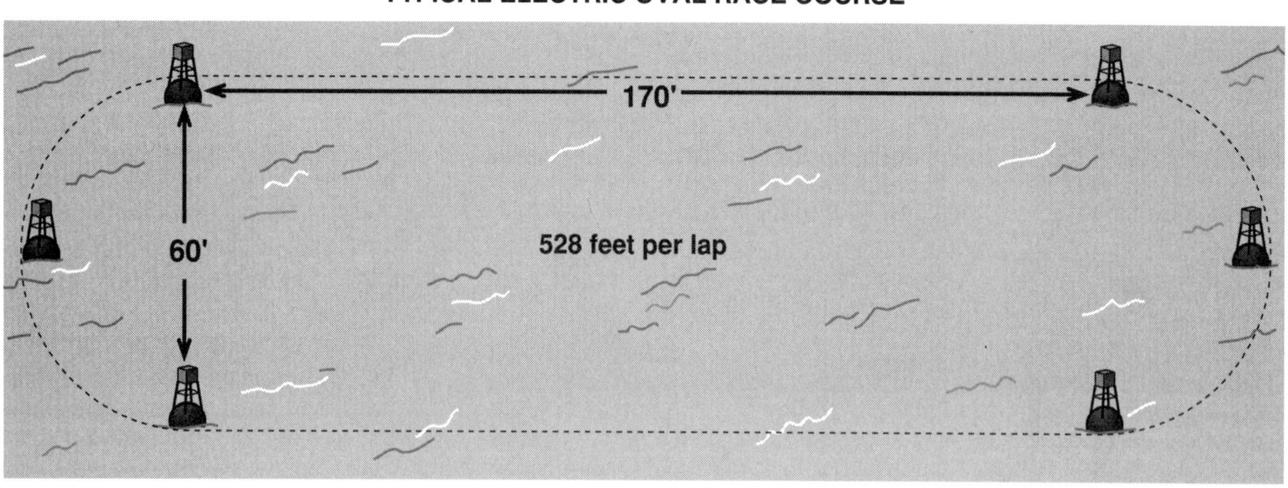

IMPBA — 4 laps = 4/10-mile heats

APBA — 5 laps = 1/2-mile heats

A typical fast electric outdoor oval, as raced by the IMPBA and APBA. Over the years, the course was lengthened substantially to cope with increased boat speeds. A typical heat lasts anywhere from 60 seconds for a four-lap hydro heat to around 2 minutes for a five-lap Crackerbox heat.

tors, even though most of them race in almost every event.

WHY RACE?

I think folks race mainly because they enjoy participating with a group of people who enjoy running R/C boats. The racing may be of secondary importance to the friendships made, and the technical information members share can't be found in books. The problems you may be having with your new boat are probably old hat to some other members, and they may be able to save you a lot of time and trouble—not to mention money—by sharing their experiences with you.

Of course, some of us just have to see whether our boat really is faster than the other guy's, and the best way to find out is through organized racing. Rather than start from scratch when organizing a club, we chose an existing national organization and built on their expertise. Boat classes, courses, scoring, rules, etc., have been worked out over the years by others, and by adopting them, we saved much time and frustration. Though races may not always run quite like clockwork, they're fun and as fair to all competitors as possible. The result is a good competitive format that tests building, setup and driving skills.

The pits are busy; racers repair boats and get ready for their next heat. If, during a race, you have an organized pit station with good tools and plenty of spare parts, you'll be less stressed.

tough to beat.

Some racing classes are divided even further according to motor types:

- open or modified motors;
- stock or closed-endbell types.

Others, such as the 6-cell Crackerbox class, require ROAR stock motors. Check to see which is most popular in your area, but be aware that the boats in some "stock" classes aren't much slower than those running open motors.

Of course, hydros and monos usually race in different classes, too, and some clubs give outboard tunnelhulls their own class. Special one-design classes, such as the Crackerbox and Key West classes, were developed to ensure that races would test driving skills rather than boat speed, and they're a good place to start a racing career.

CLASSES

All national racing organizations group boats into classes according to their expected performance; and cell count, or the voltage available for the motor, is used to predict this. In most cases the higher the voltage, the faster the boat, so to help equalize the racing, models are grouped in classes according to cell count.

The most popular classes run seven and eight cells; this is because of the availability of ready-made 7- and 8-cell packs for R/C cars, the moderate size and cost of the boats and the high speeds possible. Seven-cell hydros can exceed 30mph, and high-performance monos easily top 20mph. Though 8-cell boats have the potential to reach higher speeds, in reality, there's often very little difference in speed between 8-cell and 7-cell boats.

Other popular classes are 6-cell, 12-cell and more than 12 cells (often divided into smaller groups). You might think that 18 and 24 cells would produce substantially higher speeds, but that isn't the case. The fastest racers usually run 12 cells. More cells means more weight and a larger boat, and these reduce a boat's potential for speed. Straightaway record holders may be faster with 18 cells or more, but on a closed course, the 12-cell boats are

RULES AND ORGANIZATIONS

There are three major racing organizations in this country:

- **International Model Power Boat Association (IMPBA)**—the oldest one;
- **North American Model Boat Association (NAMBA);**
- **American Power Boat Association (APBA)**—the youngest.

All three provide rules, support, insurance and other necessities for individual clubs, both glow and electric.

The organizations' rules differ in detail, but are similar overall. Though cell counts and class names differ, courses are similar, as are the

On the elevated drivers' stand, these racers wait for the command to launch their boats. When there's glare off the water, polarized glasses help you to keep your boat in view. To avoid radio interference, be sure to keep your transmitter antenna away from those nearby.

general racing rules. If you're considering forming a local club and aren't sure of which organization to join, get information packets from every one and compare them. Generally though, because the organizations are somewhat regional, it's best to go with the one that's most active in your area.

Of course, informal racing is a lot of fun, too, and if there's no club in your area, you may want to get together with a few other boaters and set up your own racing rules. You need only a simple course, and a few bleach bottles anchored with string work well as buoys. Plan to run a certain number of laps, and make the first boat across the finish line the winner. Alternatively, count the total number of laps run in 3 or 4 minutes. This requires a balance between speed and endurance, and top speed won't be the deciding factor.

RACE PREPARATION

Some boaters say that if something can go wrong with a boat, it will go wrong in a race. This is true to a surprising degree, and a major part of preparing for a race is eliminating as much risk of mishap as possible. More new racers lose points and races because their boats failed than for any other reason; a slower boat that never breaks down will finish the season ahead of a faster but less consistent boat. Check your boat out carefully days before the race, and correct everything that goes wrong. You'll have enough to worry about on race day without having to think about that frayed flex cable breaking, or that bad solder joint on the battery pack.

Your boat should be almost ready to run before you get to the race site. The motor and drive line should be fully lubricated and assembled, the batteries fully charged and the transmitter batteries checked and replaced as needed. The only thing you should have to do for the first heat is to turn on the switches, seal up the hull and drop the boat into the water.

Part of race preparation is taking along a pit helper to help launch and retrieve your model. This helper will also serve as your co-driver to help keep you out of trouble on the racecourse. You can usually "borrow" someone at the race site, but it's better to take your own crew. This way, you'll have experienced help who understands your driving style and whom you can trust to give you reliable assistance.

DRIVING IN A RACE

Good driving is one of the most difficult things to learn in fast electric boating. No one is born knowing how to drive well, and the only way to become good is by practicing on the course with other boats. Boaters

In some events, before the racing, there's a "concours" competition in which models are judged according to how well they have been built. These beautiful models are not only pretty, but they're also very fast!

who have raced R/C cars have a definite advantage here, because their experience is almost directly transferable to boating. Though most new boaters want to start winning right away, this rarely happens. Even the fastest model boat must be driven well to win consistently.

Depending on your own learning curve and on the level of competition at the local club, it may be a season or two before you begin to realize some success at the trophy stand. Don't be discouraged; just continue to pile up the racing practice and continually try to improve. Even if you don't *place* any better by the end of the first season, you should notice an improvement in your driving. Perhaps your starts will be less disorganized, or you'll be holding a tighter line around the buoys. You may be comfortable now when your boat is alongside another boat, carving a path around the turns. Perhaps you made a particularly good pass in the last heat. All of these things—and more—are indications that your driving skills are improving.

For the flying start that's popular for fast electric racing today, it's important to position your model just behind the starting line when the command "Go!" is given; the boat should be at full speed at this time, too. Practice before the race to determine when you need to leave the bank to arrive at the starting line at the right moment. If the starting line looks too crowded, it may be wise to lay back a few yards and let the boats in front fight it out. Many heats are marred by crashes that happen between the start line and the first turn, and if you're slightly behind the wrecks, you can avoid them and end up toward the front or in the lead.

The rules of the race's sanctioning organization will affect your racing strategy, particularly regarding the buoys. At IMPBA races, cutting a buoy incurs a 50-point penalty that you may be able to justify incurring

When the last boat has been pulled off the water, there remains the suspense of the awards presentation. This lucky racer is receiving a hard-won trophy, but at a fast electric race, all competitors are "winners."

This beautiful CESA monoplane won the concours event at a local regatta. Concours judging recognizes a modeler's building skills rather than his racing ability. Some competitors are lucky enough to win both the race and concours at one event.

if cutting the buoy nets you first place. Under APBA rules, that buoy cut would cost you a one-lap penalty that you probably wouldn't be able to make up, so at an APBA race, it pays to drive a little farther off the buoys.

Most passing occurs in the turns, so if you can hang back just to the outside of the lead boat as he enters the turn, you may be able to slip inside if he drifts a bit wide.

By the same token, if you're leading, don't give your opponent a chance to pass on the inside by taking a turn too wide.

Detailed racing strategy is beyond the scope of this book, but if you want to learn more, check out John Finch's book, "R/C Powerboat Racing," which is published by Air Age Publishing.

It isn't surprising that with the high speeds today's fast electric boats can attain, racing strategy is very similar to that used with glow-powered models. With a fast, dependable boat and good driving, you, too, can visit the winners' circle.

National Organizations—1995

IMPBA, 25442 Powell Rd., Brooksville, FL 34602.
Contact Sue Bathauer, Executive Secretary: (904) 754-9030.

NAMBA, 1815 Halley St., San Diego, CA 92154.
Contact Cathie Gailbraith, Executive Secretary: (619) 424-6380.

APBA, 5001 Stormy Ct., Bakersfield, CA 93309.
Contact Wally Stewart, Chairman: (805) 398-1266.

These happy racers enjoy the fruits of their labors. National races are appealing because you get the chance to compete against top racers.

CHAPTER 14

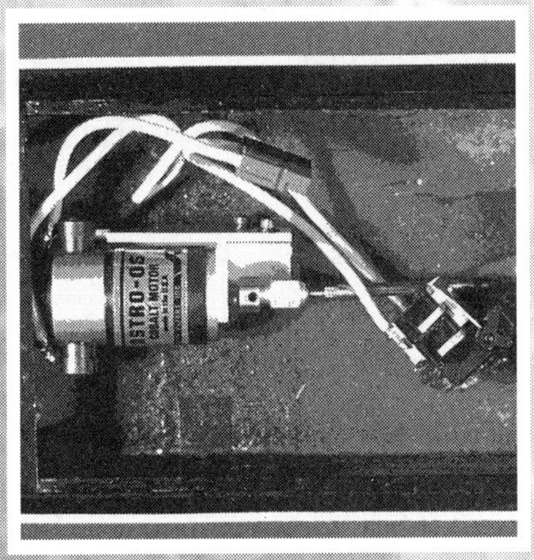

Scratch-Build a Fast Electric

14 SCRATCH-BUILD THE XS-CAT

The information presented in this chapter was first published in the July '92 issue of Radio Control Boat Modeler.

Although I've built many wooden and fiberglass R/C boats out of kits, I've always enjoyed scratch-building from my own plans. A few of my earlier designs have wooden frames and planking, but foam is now my primary building material. It's easy to work with, and foam boats don't sink! In this chapter, I describe the design and setup of one of my efforts.

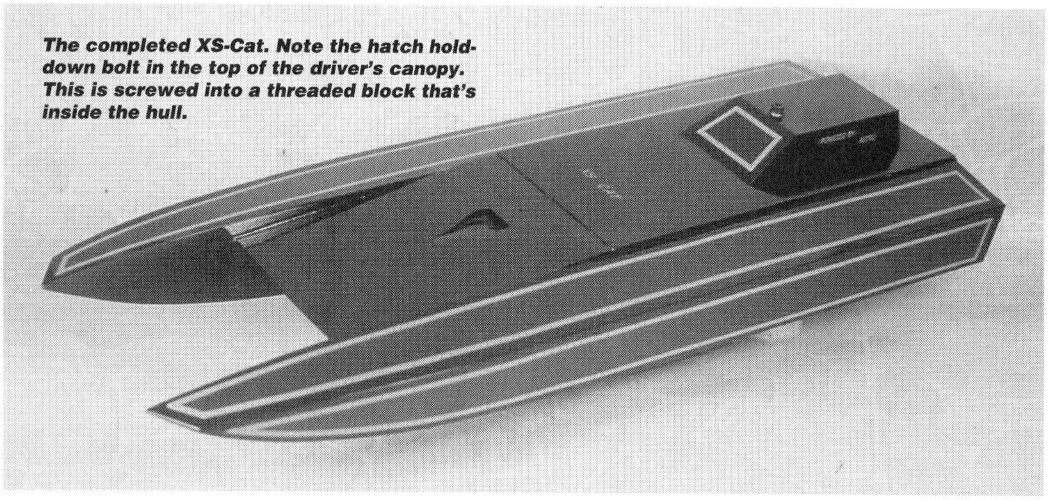

The completed XS-Cat. Note the hatch hold-down bolt in the top of the driver's canopy. This is screwed into a threaded block that's inside the hull.

CATAMARAN HULL DESIGN

The original full-size catamarans (cats) were developed for offshore racing. Designers cut the deep-vee hull in half lengthwise and installed a platform between the two "sponsons" that were formed. This creates an air-trapping tunnel that provides the catamaran with lift that raises the sponsons out of the water and increases the boat's speed. The sponsons have a large deadrise that allows the boat to operate more smoothly in rough water. Catamarans are so fast and stable that deep-vees can seldom compete successfully against them.

R/C model cats have these performance advantages, too. In fact, they're so fast that they must often compete against outrigger hydroplanes. Most successful 05-powered models are about 2 feet long. Although shorter boats are usually lighter (and, therefore, faster), I wanted to build one that would handle rough water. (Alaska's summers are rather short, so I often have to run in marginal weather if I want to run at all.) I carefully studied several successful model cats, and then I designed plans and patterns for the 25x9-inch XS-Cat. The plans are available from Air Age Inc.

CONSTRUCTION

First build the hull's center section. The bulkheads and the transom are made of 3/16-inch-thick balsa sheet, and the upper and lower surfaces are of 1/32-inch-thick plywood. The lower plywood skin runs from the nose to the transom; the upper one stops at the break in the upper surface, about 7 inches aft of the nose.

Reinforce the nose area with a balsa filler block that has been sanded to shape. To strengthen the covering and give the deck opening extra support, use 3/16-inch-square balsa stock where indicated on the plans. Reinforce the motor and battery compartments with a 3/16-inch-thick balsa floor piece that's long enough to run from the area underneath the

SPECIFICATIONS

Type: electric racing catamaran
Length: 25 in.
Width: 9 in.
Tunnel: 5 in. wide; 0.75 in. deep at transom
Weight: 3 lb., 10 oz. (with 8 cells)
Motor: Astro Flight 05 Cobalt Marine
Prop: Octura X432, X435
Duration: 2.5 min. at full speed
Amp draw: 29
Power: 278 watts (8 cells); 348 watts (10 cells).
Speed: 25.1mph (8 cells); 29.9mph (10 cells).
Radio: Futaba* with S-28 steering servo; homemade on/off switch with two 15A microswitches and a microservo.

Features: made of wood and foam, the XS-Cat is easy to build and can be powered by many different motors. Its removable hatch makes motor maintenance and battery replacement a snap. The patterns and plans are full-size, and 1/32-inch plywood is used throughout the construction. A flex-cable, surface-drive system is used, and there's enough room for up to 10 cells.

Comments: the sponsons have a wedge-bottom design that increases lift and, thus, speed. The wedges are easy to build and are detailed in the plans. For maximum speed and stability, keep all the edges below the water line as sharp as possible. Although the model could be built shorter to save weight (and increase speed), the length shown in the plans makes it an excellent rough-water performer.

motor to the transom. Waterproof the floor piece with resin and light fiberglass. Glue ½-inch tri-stock balsa along the edges of the radio-compartment opening, and make the lid out of a thin piece of Lexan.

Be sure to build the center section square and true; you must be able to align it with the sponsons, which will be attached to its sides. (The sides must form 90-degree angles with the upper and lower coverings.) Also pay attention to the direction of the grain that's called for in the plans.

Build the 1⁄16-inch-thick-plywood upper deck cowl separately from the hull, and glue ⅛-inch-square basswood strips along its edges to strengthen it. The cowl runs from the break in the upper skin to the end of the transom, where it's cut flush. Add a plywood strip to the bottom front edge of the deck opening so that the lower part of the cowl is flush with the upper deck.

Cut a NACA air scoop into the upper deck. To allow the cooling air to escape, leave the rear of the cabin open and cut openings under it. Alternatively, leave off the NACA scoop, and make the hatch solid to keep out water. Water-cooling can be used to prevent the motor from overheating.

The cabin and the upper cowl are simple, so they lend themselves to creative modification. Just keep it light!

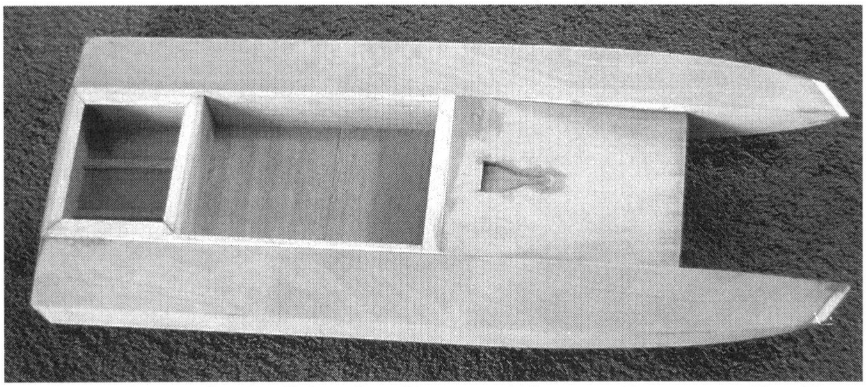

In this top view, notice the light-colored basswood pieces on the sponson tips. These strengthen them so that they won't be damaged in a crash.

SPONSONS

I recommend that you make the sponson cores out of 2-inch-thick, blue, construction-grade Styrofoam, which is very easy to cut with a band saw or a hot-wire foam cutter. Make all the surfaces of 1⁄32-inch-thick plywood. The plans include full-size templates that you'll use when shaping *and* covering the finished sponsons.

Cut the foam and the templates so that they're slightly larger than they have to be. (You'll sand them to fit later.) To ensure that both sides will be identical, stack all the similar template parts together and sand them.

Glue the inside template to the foam block, and trim the foam so that it's even with the rear of the template. Glue the aft plywood template into place, and cut the sponson top and bottom to shape using the *inside* template as a guide. (The angle of the bottom is the same from the transom piece forward.) Shape the sponson's outside edge using the top template as a guide. When you've cut the sides to shape, glue the top piece onto the sponson so that it overlaps the rear, inside plywood edges. When the glue has set, block-sand the excess material until it's flush with the edge of the sponson side.

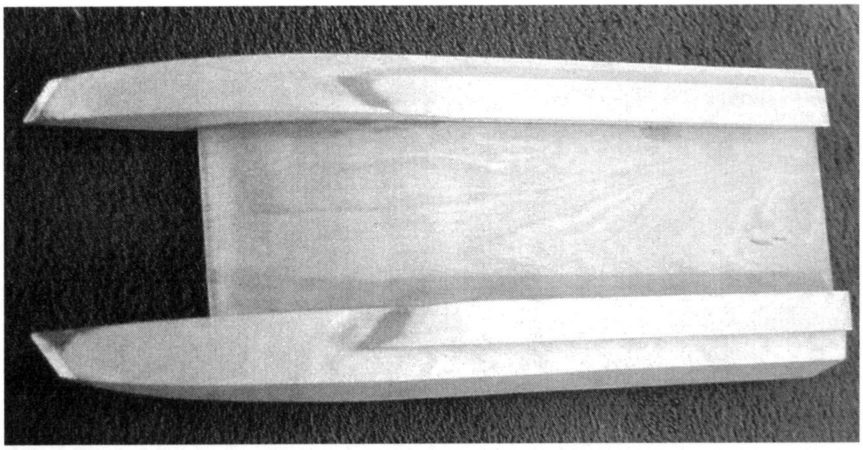

This bottom view of the XS-Cat shows the length and shape of the sponson wedges.

JOINING THE HULL PIECES

It's easier to sand the three main hull pieces before you join them. Make a jig, and glue each sponson to the hull center section. (Be sure to maintain the ¾-inch tunnel depth.) For this step, it's best to use 15-minute epoxy so that you have enough time to align everything properly. Also, to ensure that both sponsons are positioned at the proper angle relative to the inner hull's angle of attack, assemble everything on a flat surface.

I wanted the bottom of my boat to be sophisticated, so I borrowed from the technology used on some full-size hydros and put a "wedge" on the bottom of each sponson. The flat surfaces of the wedges generate extra lift, and that increases the boat's speed. The wedge's sharp edges also improve the boat's performance in the turns.

In these views of the transom, the Octura strut in the aluminum angle bracket and the MRP metal rudder are visible. To adjust the prop's depth and angle, just loosen the top bolt on the strut. Use an X432 or an X435 with seven or eight cells.

Building the wedges is easy: just glue balsa aileron stock onto the sponson bottoms and cover it with 1⁄32-inch-thick plywood. Because these surfaces greatly affect the boat's top speed and handling, they must be absolutely straight and flat, so carefully block-sand them to perfection.

I extended the plywood covering on the wedges into the tunnel area by about 1⁄8 inch (see photos). This setup causes the water to be thrown clear of the tunnel walls, and this reduces friction and makes the boat a little faster on the straights. It also allows the tunnel edges to "grab" the water well in the turns. In another attempt to reduce drag and increase speed, I made all the edges that contact the water as sharp as possible.

FINISHING

To cover or finish the XS-Cat, you can use any method you're comfortable with; just remember to keep it light! Sand everything with 220-grit sandpaper, fill dings and dents with Model Magic filler and let it dry. Apply four light coats of polyurethane spray paint at intervals of about 10 minutes, and then apply a heavier coat to eliminate the "orange-peel" look. You don't have to prime before you use this paint because it has excellent filling properties and, when it has dried, you'll still be able to see some of the wood grain. Let it dry for 24 hours, wet-sand the boat with 400-grit sandpaper, and repeat the process until the hull has a smooth, shiny finish.

DRIVE TRAIN

I power my model with Astro Flight's 6-turn Marine Cobalt 05. It's based on Astro's 05 FAI motor, which was used to set many electric boat speed records. (For more information on this, see the article on Jim Auguston's boats in the September '91 issue of *Radio Control Boat Modeler.*) The Astro Cobalt motors are still quite popular in modified racing classes, and they require less maintenance than most high-performance R/C car motors.

I use Octura driveline parts (including an aluminum strut) and an MRP metal rudder. I inserted the 0.098-inch flexible cable into a Teflon-lined brass tube, which I ran through the hull bottom and into the strut's bearing housing. If you use a motor with ball bearings, you won't need a separate thrust bearing in this drive system. If, however, you run a motor with bushings, be sure to use a thrust bearing such as the one shown in "Marine Tips" in the Fall '90 issue of *Radio Control Boat Modeler.*

Initially, I positioned the bottom of the strut housing so that it was even with the bottom of the wedge surfaces and parallel to the sponson bottom. To reduce the chance of the boat's spinning out, I limited the rudder's range of motion by attaching the pushrod to the innermost hole in the servo arm and the outermost hole in the rudder arm.

PERFORMANCE

I "fueled up" with an 8-cell, twin-stick pack and headed to a pond about a mile from my house. I moved the batteries to balance the hull about 8 inches in front of the transom, and I set the XS-Cat in the water, pointing it toward the middle of the pond. At full throttle, the Cat jumped up on step in about 12 feet and quickly accelerated to a top speed of almost 20mph.

The hull seemed to run a little wet, so I brought the model in and raised the strut by positioning the strut's bearing housing so that its bottom is 1⁄16 inch above the bottom of the wedges. After this, the hull loosened up and was visibly faster in the straights. As I had hoped, the Cat cornered perfectly at top speed, and the minimum rudder movement gave me just enough control.

For the test runs, I used an Octura X432 aluminum prop; later, I tried an X430 beryllium/copper prop. With this prop, although the boat's top speed dropped only slightly, it hooked badly in right turns, and it wouldn't turn left at all. To cure this,

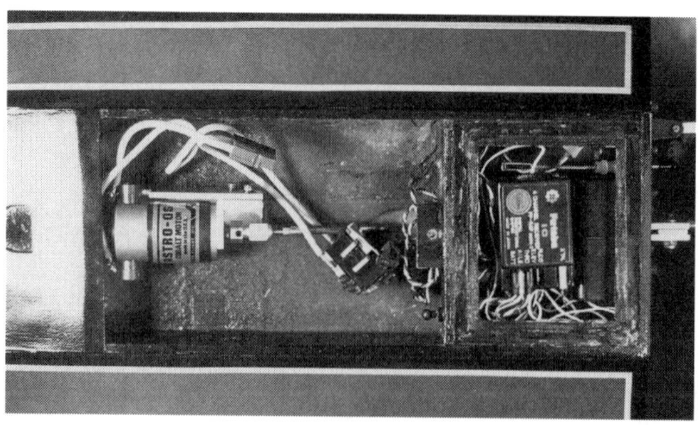

Although there isn't much room for the hardware, everything works. If I built the boat today, I'd leave out the rear bulkhead and use an ESC with BEC. This would leave much more space for installations.

I lowered the strut by 1/16 inch, and this gave the prop more bite. The boat's top speed remained high, and it took the turns smoothly and tightly.

I tested the boat with an 8-cell pack and the X432 propeller again. The motor and batteries became quite warm—but not hot—proving that the ventilation system works well. With twin 5-cell stick packs, speed increased somewhat, but I couldn't balance the boat as far aft as I wanted, and the hull ran too wet.

The XS-Cat with eight cells installed (four 1200mAh cells on each side). If you want to use twin, 6-cell battery packs and need more room, you can deepen the hull center section by 1/4 inch; be sure to maintain the 3/4-inch tunnel depth.

To raise the bow out of the water when it's running, I incorporated "steps" on the sponson bottoms. Made of 3/16-inch balsa and covered with thin Lexan, these additions give the Cat a more positive, nose-high attitude, and it runs about 25 percent faster! It doesn't take the turns quite as sharply but, by lowering the prop about 1/16 inch, I restored the former excellent handling. Even running into a brisk breeze, the Cat doesn't "take off." Then I noticed that it wasn't turning quite as sharply as before, so I increased rudder throw slightly, and the Cat regained its quick turning capability.

With 8 cells, the Cat is very quick, and with 10 cells, I've clocked it at almost 30mph. Using Astro's newest 4-turn motor, it would be even faster.

I intended to race the model, so I tuned the prop depth to maximize cornering. (Remember, the boat with the highest top speed isn't necessarily the fastest one around a racecourse.) I raised and lowered the strut and adjusted its angle and, after each adjustment, I evaluated the boat's performance by timing a few laps over a measured oval course.

The XS-Cat is a very successful, rewarding project. Why not try building one? By the way, about the boat's name—my household had four fuzzy cats before I built the model. Five is more than my wife and I needed, hence the name "XS" (excess)!

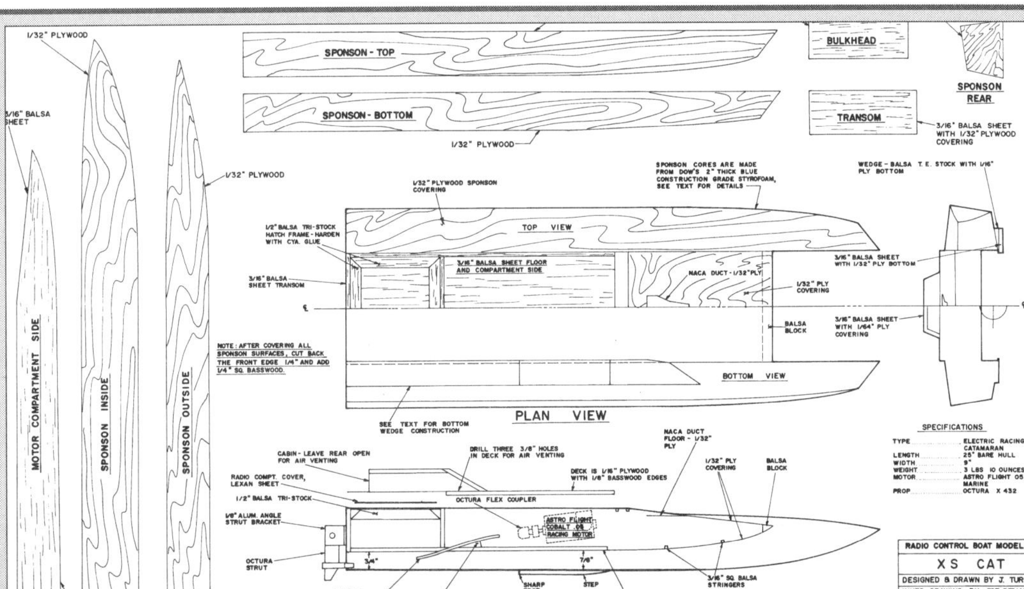

Plan no. FSPB7921 XS-Cat

Designed by Jay Turner, the XS-Cat is an easy-to-build, foam-and-wood catamaran that's perfect for an 05 motor. The well-detailed plans feature full-size templates and all the motor and hardware details. The foam sponsons are covered with 1/32-inch-thick plywood and are glued to the built-up, wooden, hull center section. The bottom of the sponsons have "wedges" to increase lift (and speed!). Length: 25"; beam: 9"; power: 05 motor; 2 channels;1 sheet. **$9.95**

To order plans for the XS-Cat, call Air Age Inc.'s mail-order department at

(800) 243-6685.

Air Age Publishing,
251 Danbury Rd.,
Wilton, CT 06897;
to order, call (800) 243-6685.

Astro Flight,
13311 Beach Ave., Marina Del Rey, 90292; (310) 821-6242.

Blackwell Racing Boats,
8655 SW Sorrento Rd., Beaverton, OR 97005; (503) 643-3961.

Bru Line Industries,
P.O. Box 3786, Center Line, MI 48015; (810) 757-0854; fax (810) 754-4080.

C&M Mfg.,
P.O. Box 701-353, West Valley, UT 84170; (801) 974-5757; fax (801) 974-0869.

Carl Goldberg Models Inc.,
4734 West Chicago Ave., Chicago, IL 60651; (312) 626-9550; fax (312) 626-9566.

Deans Connectors,
7628 Jackson St., Paramount, CA 90723; (310) 634-9401.

DPI Leisure Sports,
15500 Woodinville-Redmond NE, Building C, Ste. 100, Woodinville, WA 98072; (206) 481-2456; fax (206) 481-0719.

Dremel Tool,
P.O. Box 1468, Racine, WI 53406-1468; (414) 554-1390; fax (414) 554-7654.

Dumas,
909-AM E. 17th St., Tucson, AZ 85719; (520) 623-3742; fax (520) 620-1329.

Electric R/C,
15402 Cascadian Way, Lynwood, WA 98037.

Fantom Motors,
209 North Haver St., Hartford, MI 49057; (616) 621-2119; fax (616) 621-4619.

Futaba Corp. of America,
P.O. Box 19767, Irvine, CA 92713-9767; (714) 455-9888.

Graupner;
distributed by Hobby Lobby International, 5614 Franklin Pike Circle, Brentwood, TN 37027; (615) 373-1444; fax (615) 377-6948.

Horizon Hobby Distributors,
4105 Fieldstone Rd., Champaign, IL 61821; (217) 355-9511.

Hughey Boats,
P.O. Box 68328, Indianapolis, IN 46268.

K&B Mfg.,
2100 College Dr., Lake Havasu City, AZ 86403; (602) 453-3579.

Kyosho;
distributed by Great Planes Model Distributor, P.O. Box 9021, Champaign, IL 61826-9021; (217) 398-3630; fax (217) 398-0008.

Litespeed,
P.O. Box 4765, Spokane, WA 99202; (509) 535-2717.

MRP,
1594 S. Anaheim Blvd., Ste. B, Anaheim, CA 92805; (714) 491-8519; fax (714) 758-0150.

Novak Electronics,
18910 Teller Ave., Irvine, CA 92715; (714) 833-8873; fax (714) 833-1631.

Octura Models,
7351 N. Hamlin Ave., Skokie, IL 60076; (708) 674-7351; fax (708) 674-7363.

Parma International Inc.,
13927 Progress Pkwy., North Royalton, OH 44133; (216) 237-8650; fax (216) 237-6333.

Prather Products,
1660 Ravenna Ave., Wilmington, CA 90744-1398; (310) 835-4764; fax (310) 835-0810.

Sanyo Energy Corp.,
2001 Sanyo Ave., San Diego, CA 92173; (619) 661-6620; fax (619) 661-6743.

Sermos RC Snap Connectors Inc.,
Cedar Corners Station, Box 16787, Stamford, CT 06905; (203) 322-6294.

Sterling Models,
3100 "C" St., Philadelphia, PA 19134.

Tekin Electronics,
940 Calle Negocio, San Clemente, CA 92673; (714) 498-9518; fax (714) 498-6339.

Trinity Mfg.,
P.O. Box 1519, Hamlet, NC 28345; (800) 632-6328.

Twister Motors,
657 E. Arrow Hwy., Ste. H, Glendora, CA 91740.